A CHANGE OF HEART

A CHANGE OF HEART

*How the Framingham Heart Study Helped Unravel
the Mysteries of Cardiovascular Disease*

Daniel Levy, M.D.
& Susan Brink

Alfred A. Knopf　New York 2005

Library of Congress Cataloging-in-Publication Data
Levy, Daniel, and Susan Brink
A change of heart : how the Framingham heart study
helped unravel the mysteries of cardiovascular disease /
Daniel Levy and Susan Brink.—1st ed.
p.   cm.
ISBN 0-375-41275-1
1. Cardiovasular system—Diseases—Massachusetts—Framingham—
Epidemiology—History—20th century. 2. Health risk assessment—
History—20th century. I. Brink, Susan II. Title.
RA645.C34L48 2004
614.5'91'0097444—dc22   2004048589

Manufactured in the United States of America
First Edition

To the people of Framingham.
The world owes you an enormous debt of gratitude.

To my family for encouraging me throughout this long journey.
—Daniel Levy

To my mother, Sophia; and my daughters,
Jenny and Rachel: the women who fill my heart.
—Susan Brink

CONTENTS

ACKNOWLEDGMENTS

This book would not have been possible without the more than fifty years of dedication and commitment from three generations of Framingham Heart Study volunteers. We would like to thank them all for providing a gift to the world that has changed untold millions of lives. In particular, we are deeply grateful to William Sullivan, Victor Galvini, and Evelyn Langley, who spent hours with us recalling the early days of the Study. We have drawn on the cardiovascular disease research of dedicated scientists going back more than a century. We acknowledge with gratitude the thoughtful recollections of former Framingham Heart Study directors Thomas Royal ("Roy") Dawber, William Kannel, and William Castelli. Patricia McNamara, the data manager who devoted more than three decades of her life to the Study, provided us with her keen recollections in the days before her death. In addition, we would like to thank Edward Freis, William Zukel, Henry Blackburn, Jeremiah Stamler, Manning Feinleib, Sandra Savage, Michael Brown, Kilmer McCully, Ralph Paffenbarger, and Arthur Caplan for their insights. We thank Paul Sorlie of the National Heart, Lung, and Blood Institute for providing access to archival material on the early years of the Framingham Heart Study; Gerald Oppenheimer for adding valuable historical perspective on the launch of the Study; the staff of the Countway Library at Harvard Medical School for finding and providing copies of previously published correspondence between Paul Dudley White and President Richard Nixon; and the library staff of *U.S. News & World Report* for tracking down countless medical studies and books. We are appreciative also of assistance from the Framingham Historical Society. Finally, we would like to thank, for their thoughtful critiques and insightful suggestions, Susan Levy, Zorba Paster, William Kannel, Greta Lee Splansky, June Davidek, Susan Baum, Marjorie Howard, Linda Hosek, and Lance Dickie.

A Change of Heart

INTRODUCTION

Today everyone knows that smoking cigarettes poses a health risk for cardiovascular disease. It seems we have always realized that a high-fat, high-cholesterol diet coupled with a sedentary lifestyle is a recipe for a heart attack. Who now doesn't know his or her blood pressure and cholesterol level, and that high numbers indicate increased risk and necessary treatment? And we all are aware that obesity is a health problem of epidemic proportions, not just an aesthetic consideration.

For the 78 percent of Americans who are under fifty-five years of age, such lifestyle information is, indeed, old news.[1] For them, these facts have unfolded steadily throughout their lifetimes. They've grown up with them, heard them for as long as they can remember. So ingrained is the information in the national consciousness that many may well be unaware that during the last half century, science has made tremendous progress in understanding the causes of heart disease and in developing methods to prevent it.

The quest to understand the causes of heart disease has resulted in nothing short of a quiet revolution. The strides toward prevention and treatment have been taken in the slow and tedious realm of medical research. What we have learned about cardiovascular disease has changed the focus of medicine, from treating disease after it develops (secondary prevention) to preventing it before it takes hold (primary prevention) to deterring the emergence of risk factors that promote disease (primordial prevention). This wealth of medical information has altered the eating and exercise habits of millions, reduced greatly the number of smokers, and prompted millions to seek regular medical checkups even when they feel well. Best of all, our taken-for-granted understanding of how cardiovascular disease develops has contributed to a steady decline in annual death rates from heart problems. Since its peak in 1963, the death rate for coronary heart disease has fallen 60 percent, that for stroke 66 percent.[2]

A turning point in our evolving understanding of heart disease was the establishment of the Framingham Heart Study in 1948. It was a large and ambitious community-based research project unlike anything that had been conducted before. It came at a time of growing awareness that cardiovascular disease was sweeping the country, even slowing down what should have been a steady rise in life expectancy.[3] It was also a time, three years after the end of World War II, when resources from the national treasury, no longer needed for military purposes, could be used for research into the nation's leading killer.

A trip back in time to that era would shock Generation Xers. They would immediately see that the meals then considered healthy were full of saturated fat and cholesterol. They would watch young and vigorous men and women adopt habits that removed walking and other forms of exercise from their lives. And they would squint at it all through clouds of cigarette smoke, whether in restaurants, offices, homes, elevators, trains, or airplanes.

Coronary disease was an epidemic. It was cutting lives short with such methodical regularity that most Americans in 1948 regarded early death from heart damage as an unavoidable act of fate. The epidemic respected neither money nor power. Once it struck, medicine could offer no treatment and scant hope. Physicians were so baffled by the cardiovascular system that they often didn't even know what was killing their patients.

In light of this ignorance, the U.S. government in 1948 made a twenty-year commitment to uncovering the root causes of heart disease. That scientific resolve was sponsored by the U.S. Public Health Service with half a million dollars of start-up funding from Congress. A cadre of physicians, scientists, government officials, and academics— many of whom knew each other from having served together at military hospitals during the war—selected a New England town in which to carry out this national scientific experiment.[4] The Framingham Heart Study turned out to be instrumental in changing the attitudes, if not the behavior, of virtually every American, and it put the otherwise ordinary town of Framingham, Massachusetts, on the map.

The two-decades-long commitment grew into a project now in its sixth decade. Framingham research has helped Americans understand the vast difference between what is an average state of health and what is desirable. Average blood pressure is undesirable. Average cholesterol levels in the United States are unacceptably high. Average weight, we now know, is too fat. Average diets are too high in saturated fat and cholesterol, and average exercise habits are too sedentary.

By looking at the way ordinary people from Framingham lived and died, researchers discovered critical information on the natural history of coronary disease. But they went further. They added the term "risk factor" to the lexicon, and proved that people can do something about the things that put them at risk. Today we know a lot about who is at risk for cardiovascular disease before anyone ever suffers a life-threatening event.

Indeed, the things that put average citizens at risk are so ingrained in the American psyche that it's hard to remember a time when people didn't know them. Yet in 1948, people had no idea how much they didn't know. By the time I visited the Framingham Heart Study for the first time in 1982, the original Study participants had been coming in every two years to give their medical histories and get physical examinations, electrocardiograms, chest X-rays, and a variety of blood tests. Since 1979, echocardiograms and exercise treadmill tests were part of the exam.

On a research rotation during my medical residency at Boston University Medical Center, William Castelli, the Study's director, asked me to undergo the same examination as the participants. It was my initiation rite, my first exposure to the voluntary half-day exam that participants experience every other year. I went early in the morning, had a fasting blood sample taken, walked on a treadmill, blew into a bellows to have my lung capacity measured. I had six spots on my chest gently sandpapered where electrodes were attached for an electrocardiogram. I had a physical exam and gave a detailed medical history. It was the beginning of three months at the Study that would focus my career and change my life.

As far back as my medical school years at Boston University, I had been interested in cardiology because the field was full of promise and challenge. Angioplasty was becoming a popular alternative to bypass surgery as a way of opening clogged coronary arteries, and clinical trials were beginning to sort out the relative value of the two treatments, comparing them also to medical therapy. Echocardiography was just taking off as a means of imaging the heart and diagnosing a wide range of defects and measuring the damage done by heart attacks.

At the same time, trials were testing the effectiveness of lowering blood levels of cholesterol and controlling hypertension. As a medical resident I worked closely with Daniel Savage, the first cardiologist hired by the study. When the Study began, there were only a few cardiologists in the country, and even the first directors were generalists who came up through the ranks of the Public Health Service. Savage was instrumental, along with Castelli, in bringing new cardiology tests to Framingham, including echocardiography, ambulatory electrocardiography, and treadmill testing.

Framingham provided what I had been looking for: a research opportunity combining cardiology and epidemiology. Even twenty years ago, it was a rare career path for a cardiologist. Many trainees pursued careers in invasive cardiology, lured in part by the technology, the heroics, and the instant gratification that come with performing balloon angioplasties and stent insertions to open clogged coronary arteries. Today, we are getting some of the best and brightest as fellows at the Study, coming from the leading medical schools in the United States, Canada, and overseas. Our fellowship program now offers training in that combination of disciplines, and many of our fellows enroll in epidemiology and biostatistics course work at Harvard or Boston University and obtain advanced degrees in public health.

Preventive cardiology was Framingham's contribution to the world, and I was drawn to it. I read old research papers going back to the Study's beginning and saw how data were collected and analyzed. I was aware of how much good had already come from it, and how much was left to discover. I noted the connection between recent clinical trials with no direct relationship to the Heart Study yet clearly extending its findings by proving the value of lowering blood pressure. Framingham results were pinpointing hypertension as a risk factor for heart disease and stroke. Cholesterol-lowering trial results were also starting to weigh in, as other realms of science began confirming Framingham findings on the hazards of high levels. The data from the Study were starting to change people's lives. By 1984 I was in the middle of a cardiology fellowship at Harvard's Brigham and Women's Hospital in Boston and I was eager to spend a few years doing research. Just then, Savage was leaving the Framingham Heart Study to take a position at the National Center for Health Statistics. Castelli called to tell me he was looking for a cardiologist to replace him. "Please take some time to consider the job." It was an offer I couldn't refuse. I thought I'd try my hand at research for five years and then reassess. But the challenge and excitement of asking questions and digging into a treasure trove of Framingham data to find the answers have continued to enthrall me.

Fourteen years later, I was again poring over research papers from the Study. This time, I was compiling landmark Framingham publications for a book commemorating the Study's fiftieth anniversary. I reread the very first papers written in the late 1940s describing intentions and methods. I perused the classic paper by William Kannel and Thomas Royal Dawber (widely known as Roy), describing for the first time the risk factors for heart disease. It was this publication that introduced that term, so familiar today. I understood clearly then Framingham's central role in educating the nation about the causes of heart disease.

During its first half century, the Study's findings on the dangers of hypertension, high levels of LDL cholesterol, low levels of HDL cholesterol, smoking, and diabetes were widely recognized as major contributors to the understanding of cardiovascular disease. But the Study also contributed in more subtle ways. There was, I became convinced, a bigger story to tell. Framingham data were used to model dozens of clinical trials to test the value of such common treatments as blood-pressure- and cholesterol-lowering medications. Its results helped teach Americans how they should eat and exercise, though it could not make them follow through, and gave them one more compelling reason to stop smoking, or not start.

The Study did not act alone to uncover the answers that now save so many lives. Science is a collaboration, not an isolated effort. It is like a relay race toward answers. The baton is sometimes passed directly forward, but more often it goes in the direction that an idea or a theory sends it. The branches of science that lead to treatments and cures include laboratory studies at the molecular level, pathology studies, animal studies, and clinical trials in humans. Framingham's contribution is within the branch called epidemiology. And, as we will show in this book, its researchers refashioned cardiovascular-disease epidemiology in ways that revolutionized medicine.

This book will look at some of the personalities who figured prominently in unraveling the mysteries of heart disease, some of them actors in the Framingham Heart Study, others scientists conducting research elsewhere. We can by no means cover everyone who contributed. The Heart Study was crucial to this effort, but without the curiosity and vision of scientists going back centuries, researchers would not know where to begin. And without the work going on simultaneously in many other branches of research, its findings could not have led to breakthroughs in treatment.

"Medical breakthrough" is a dramatic phrase, but most improvements come from the persistent efforts of hundreds of disparate research projects. Many batons are carried and juggled simultaneously as some hypotheses are proved and built on and others discarded. Those who devote their careers to disproving a theory still aid the ultimate cause of science by showing that a once promising path needs reassessment.

The researchers started out with few assumptions about the underlying causes of heart disease. The scope of their study seems modest by current standards: monitor healthy people and follow them for the development of cardiovascular disease. The clinical traits of those who developed the disease were studied as possible risk factors. Once potentially causal relationships were identified and confirmed, other branches of science could test novel treatments and strategies for prevention.

When factors such as high blood pressure and cholesterol were earmarked as potential causes, researchers like Edward Freis, who believed that controlling hypertension could save lives, could prove their theories with clinical trials. Basic scientists could understand the synthesis and regulation of cholesterol, pharmacologists could discover drugs to lower it, and clinical trials could prove the drugs beneficial. The baton was passed in the other direction at times. Kilmer McCully's unshakable belief in homocysteine as a risk factor for heart disease persuaded us at the Heart Study to focus on homocysteine in our research. And our observations were part of the evidence that gave credence to McCully's controversial theory.

Discovering the cause of a given disease doesn't cure it. Information must percolate into the public consciousness. It must find its way onto physicians' prescription pads and into medicine cabinets. When it comes to prevention, information about the causes of heart disease must be disseminated to the public so that individuals can adjust the way they eat and live. The Framingham Heart Study showed doctors why today one in two American men will eventually develop heart disease. When the Study began, medical science believed that women were immune to the affliction. Framingham researchers included women in their Study not out of chivalry or a commitment to equality, but rather to observe what it was that was protecting women so they might use the insights to help men. Yet researchers there would discover that women were also susceptible and that one out of three will develop the condition during her lifetime.

When the Study began in 1948, medical science didn't know that high blood pressure, a common finding, was a powerful cardiovascular-disease risk factor. Doctors were trained to believe that a systolic pressure of 100 plus one's age was healthy—a belief that starts being dangerous at the age of forty and can be fatal to a seventy-year-old. Fifty years of measuring blood pressure on citizens who volunteered to take part in the Study and were followed for the development of cardiovascular disease taught the rest of us that high blood pressure is not benign, but rather a dangerous contributor to heart failure and stroke. The people of Framingham showed that heart attacks can come on silently, and hundreds of thousands of death certificates, which once bore the cause of death as "acute indigestion" or plain "unknown," could now be relabeled "coronary heart disease." The smokers of Framingham, who died years or decades too soon, taught us that tobacco increased their odds of cardiovascular problems—not to mention lung cancer—and that filtered cigarettes offered no protection. Framingham volunteers demonstrated that being overweight increases the chances of developing diabetes, hypertension, and heart failure, and those who began exercising showed us that we have the power to reduce those risks.

What researchers discovered about the riddle of heart disease today sounds too easy. Eat a healthy diet. Exercise. Don't smoke. Know your blood pressure and cholesterol levels, and change habits and use medications if necessary to control them. The answers are often inexpensive, accessible to the rich and the poor. They are pure common sense. But it wasn't always so obvious.

The Heart Study researchers didn't go it alone. They needed the 5,209 men and women from Framingham at first, followed by 5,124 of their sons and daughters, and now 3,500 of their grandchildren who have donated their medical histories to science. It is ironic, perhaps, that this most respected—even beloved—piece of epidemiology centers on the heart, the organ that symbolically aches, breaks, longs, and loves like no other. It took a commitment from thousands of volunteers to make the study a success.

Researchers began their work at a time when Americans were less mobile. The original Study volunteers grew up in Framingham and stayed there to raise families. But we don't stay put like that anymore. Framingham residents, like people across the country, have moved far and wide. Yet their devotion to the Study keeps them coming back for follow-up exams, and keeps their children and grandchildren coming back, to continue to provide clues to the underlying causes of heart disease.

They are typical Americans, as thin, active, overweight, or sedentary as Americans anywhere. Their interest in this Study was aroused in 1948, and nurtured by its directors for decades. But over time, they realized they were making their contribution not for their own benefit, or even that of their families, but for all of humankind. In an experiment little known to outsiders, one New England town changed the practice of medicine and the lifestyles of tens of millions.

ONE

A Killer of Paupers and Presidents

It was April 12, 1945, and the country was heartbroken. Franklin D. Roosevelt, the thirty-second president of the United States, died suddenly in what had come to be known as the Little White House, a cottage in the woods of Pine Mountain near Warm Springs, Georgia. The public was unprepared for his death, though for many months his doctors knew that he was gravely ill. In keeping with the culture of the times, his personal physicians hid the grim reality of the president's failing health from the press, from the public, from his family—even from FDR himself. A casualty of an as yet unrecognized epidemic, the leader of the free world slipped away.

Roosevelt, his doctors, and the media had colluded to portray him as the picture of health. Long before he was elected president, in the summer of 1921 when he was thirty-nine years old, he fell victim to another epidemic. Polio rendered his legs nearly useless, his ability to walk nothing more than a simulation. He supported dead weight from the waist down with braces locked at the knee, and he would swing himself forward in a practiced rhythm between crutches. Throughout his life, the public saw him as strong, self-assured, and independent. No American was privy to the scene of Arthur Prettyman, FDR's personal valet, strapping full-leg braces on the president as he lay supine in bed. The metal of each brace was painted black, and the president always wore black shoes and socks so as not to draw undue attention to the contraption. It was, like the title of Hugh Gregory Gallagher's book, *FDR's Splendid Deception*.[1] His walk was seldom photographed, nor was the wheelchair on which he often depended. When a rare photographer violated the

White House rule, Secret Service agents would seize the film and expose it. Only pictures of Roosevelt in a strong, erect stance or a comfortably seated position were permitted.[2]

Rumors that Roosevelt was in poor health circulated during his first run for president and were blamed on the opposition's attempt to derail his candidacy. The country was in the throes of the Great Depression. America was mired in despair, and Roosevelt needed to prove that he was strong and steady. To still the gossip, he released his medical records in 1931.[3] His blood pressure was 140/100—the 140 systolic only marginally hypertensive, but the 100 diastolic a bad omen. Even the most brilliant medical minds of the time possessed neither the knowledge to recognize the gravity of his disease nor the tools to treat it. The numbers did not raise questions, but periodic reports continued to emerge that he was ill. So in 1932 he took out a life insurance policy for $50,000, reassuring his supporters by passing the medical examination at the age of fifty.[4] Shortly after assuming the presidency in 1933, in what may have been a fateful decision, Roosevelt selected Admiral Ross McIntire as his personal physician. Dr. McIntire was an ear, nose, and throat specialist whose main concern would be the president's numerous head colds and sinus problems.[5]

Roosevelt took the helm of a nation at a time that would have taxed the hardiest of souls. America was then home to between 13 million and 15 million unemployed workers. A couple of million of them took to the road to find employment. They created a whole class of homeless migrants. They left behind dust-ravaged farms and boarded-up factories to wander the country in search of work. Hundreds of thousands of them lived at the edge of cities in tents and shantytowns, dubbed "Hoovervilles" in disparaging reference to the president they blamed for their lot. Panic about the economy had forced the closing of banks in thirty-eight states. The plight of a stricken populace surely took its toll on their leader during his first term. "I see millions whose daily lives in city and on farm continue under conditions labeled indecent by a so-called polite society half a century ago. . . . I see one-third of a nation ill-housed, ill-clad, ill-nourished," he said in his second inaugural speech.[6] And, in words that live in memory and history, he tried to reassure Americans at his first inaugural when he said, "The only thing we have to fear is fear itself."[7]

As the strain registered in medically measurable form, McIntire hardly made note of the rise in the president's blood pressure. It was 169/98 in 1937 as Roosevelt began his second term. From then on, it would fluctuate but remain abnormally high. His vital numbers rose to 188/105 in 1941, when the Japanese bombed Pearl Harbor. Still, as is typical, he had no outward symptoms of hypertension. Roosevelt launched a nationwide war effort, committing more than 16 million U.S. troops to the Allied cause in World War II. By the time American soldiers landed in Normandy in June 1944, his blood pressure was 226/118—a life-threatening level. The limited medical technology of the day, electrocardiograms and chest X-rays, showed a damaged, enlarged heart.[8] Still, no one told FDR the bad news, nor did he ask.

Roosevelt was absent from the White House for nine weeks during the first five months of 1944.[9] In those days, he would go to Warm Springs, an impoverished farm community eighty miles southwest of Atlanta, Georgia, for an "off the record" absence from duties, which amounted to much-needed bed rest. He had gained sustenance and rejuvenation from the town's healing waters since 1924. These trips were about his only concession to poor health, and the reason behind them went unspoken. In an era when the media grant no mercy in exposing the secrets of public officials, it is difficult to fathom that back then journalists would comply with and help promote such a public deception. Dr. McIntire insisted that the president's health was good, that Roosevelt's blood pressure was normal for a man his age.[10] In his treatment notes of April 1944, when the president's blood pressure was 210/120, McIntire wrote, "A moderate degree of arteriosclerosis, although no more than normal for a man of his age."[11]

Everyone, it seems, was happy to go along with the opinion, particularly since at the time there was nothing to be done for escalating blood pressure. There is hardly an American today who doesn't know enough to shudder at the president's vital numbers. Meanwhile, McIntire remained concerned chiefly about FDR's upper respiratory system. He dosed the president daily with nose drops and sinus sprays. Containing vasoconstrictors, the drugs did little to relieve his breathing symptoms, and probably further increased his critically high blood pressure.

If the public was fooled into believing it had a healthy leader, his family was becoming alarmed at his failing appearance. His daughter Anna, who lived in the White House in 1944, became conscious of the darkening hollows under his eyes, the loss of color in his face, the soft cough that accompanied him day and night. To her observant eye, his strength seemed to be failing him; he was abnormally tired even in the morning hours; he complained of frequent headaches and had trouble sleeping at night. Sitting beside him in the movies, she noticed for the first time that his mouth hung open for long periods; joining him at his cocktail hour, she saw the convulsive shake of his hand as he tried to light his cigarette; once, as he was signing his name to a letter, he blanked out halfway through, leaving a long illegible scrawl.[12]

Careful listeners to his radio fireside chats might have noticed, certainly by 1944, an audible short-windedness that probably reflected some degree of congestive heart failure.[13] But Eleanor Roosevelt, who had little patience for the distraction of illness, attributed her husband's malaise to overwork and stress. When doctors began to urge a reduction of meat in his diet, the First Lady had prime cuts of steak delivered to the White House because her husband loved them.[14] By early 1944, however, she was ready to reject McIntire's diagnosis and ask for a second opinion.[15] It was Anna who, at last, pushed McIntire into sending the president to Bethesda Naval Hospital in March for a thorough examination.[16]

There, a young cardiologist, Dr. Howard Bruenn, pronounced the president desperately ill. But McIntire carefully controlled the disclosure of all medical information, and believed Bruenn's view of FDR's health would disturb the president and his family. In fact, he balked at Bruenn's recommended treatments, which included bed rest, a light diet, salt reduction, and a program of weight loss.[17] McIntire watered down Bruenn's suggestions until the "regimen amounted to no more than treating a cold."[18] He was even more upbeat with the public. At a press conference following Roosevelt's medical exam he declared, "I can say to you that the checkup is satisfactory. When we got through, we decided that for a man of 62, we had very little to argue about, with the exception that we have to combat the influenza plus the respiratory complications that came along afterward."[19]

Historians speculate that as Roosevelt's cardiac problems became more apparent, McIntire grew more determined to hide the reality that he had overlooked or concealed for so long. It was a reaction that "one can only assume was a protection of his turf and a desire to hide the fact that he had failed to diagnose heart problems earlier."[20] Years later, in a 1970 journal article called "Clinical Notes on the Illness and Death of President Franklin D. Roosevelt," Bruenn wrote about the frustration of treating FDR. His account of the examinations and treatments of the president was the first medical data made available apart from McIntire's memoirs. Bruenn's account contrasted sharply with the self-serving recollections of McIntire, and Bruenn concluded by saying, "I have often wondered what turn the subsequent course of history might have taken if the modern methods for the control of hypertension had been available."[21] The president's original medical chart vanished immediately after his death, and the most reliable enduring record of his health during his presidency is the notes that Bruenn kept.

Bruenn persisted in speaking his mind, calling in other experts, and eventually he prevailed over McIntire. But even with focused concern, Bruenn was virtually powerless to control FDR's severe hypertension. Roosevelt began taking digitalis, the only drug available for treatment of heart failure. At the very end of his life, he was prescribed phenobarbital, a sedative, which doctors at the time hoped would lower blood pressure. It proved ineffective. Lifestyle alterations for Roosevelt included a recommendation that he cut back on cigarettes from twenty a day to ten, but Bruenn was frustrated in his attempts to convince the president of the importance of it. Few doctors at the time considered tobacco a risk factor for cardiovascular disease, and Bruenn's concern about FDR's smoking was probably aimed at providing relief from a chronic cough and respiratory problems.[22]

He also advised the president to limit alcohol intake to one and a half cocktails a day. This may have been prescient. Alcohol in high doses can cause blood pressure to accelerate, but Bruenn could not have known that then.

For the next few months, the president rallied publicly. With his country and millions of its troops depending on his strength of command, he felt he could not quit in the middle of war, and he decided to run once again for reelection. In the year before his death, Roosevelt's blood pressure numbers through 1944, according to medical records, read like a recipe for disaster: March 27, 186/108; April 1, 200/108;

November 18, 210/112; November 27, 260/150.[23] And yet, during those months, he traveled to Hawaii to confer with top brass on military strategy in the war against Japan. He went to the Democratic National Convention in Chicago, but actually accepted the nomination for a fourth term in San Diego. He traveled to Alaska and to Washington State. He met with Winston Churchill in Canada.

Historians have shown that throughout this time he was quite ill. On one occasion, in the company of his son, James, he fell to the floor trembling with pain. Churchill even took the extreme measure of going to see Dr. McIntire because of his deep concern over Roosevelt's health. McIntire continued to insist that the president was fine. Some members of the press actively promoted the deception that Roosevelt was in robust health. Henry R. Luce, editor of *Life* magazine, sorted through pictures of the president and said, "In half of them, he was a dead man. We decided to print the ones that were the least bad."[24] While giving a speech on a ship in Bremerton, Washington, Roosevelt sounded hesitant and uncertain as he gripped the sides of the lectern for support. Hugh Gallagher writes in *FDR's Splendid Deception,* "The President's balance was uncertain; the deck of the destroyer was not stable; he gripped the [lectern], his fingers clenched with fear and apprehension. As he spoke, he felt spasms of pain radiating from his heart. He burst into a sweat, and his delivery became confused and imprecise. That great, clear tenor voice became muffled. Afterward, his doctors found he had suffered an attack of angina—a severe pain caused by a restriction of the arteries bringing blood to the heart."[25]

It was the first of two suspected public attacks of angina. The second may have occurred as he delivered his final inaugural speech on January 20, 1945. He stood to address the American people:

> Mr. Chief Justice, Mr. Vice President, my friends, you will understand and, I believe, agree with my wish that the form of this inauguration be simple and its words brief. . . . In the days and years that are to come we shall work for a just and honorable peace, a durable peace, as today we work and fight for total victory in war. We can and we will achieve such a peace. . . . I remember that my old schoolmaster, Dr. Peabody, said, in days that seemed to us then to be secure and untroubled: "Things in life will not always run smoothly. Sometimes we will

be rising toward the heights—then all will seem to reverse itself and start downward. The great fact to remember is that the trend of civilization itself is forever upward; that a line drawn through the middle of the peaks and the valleys of the centuries always has an upward trend."[26]

He delivered a message of hope, but the world was at the height of war, and the ceremony was solemn. The expense of a show of festivity would have been inappropriate, and the oath of office was taken quietly on the South Portico of the White House.

The impropriety of public celebration, ironically, served Roosevelt's failing health. The address—fewer than five hundred words—was by far his shortest inaugural speech. It was to be the last time the public would see him standing. His secretary of labor, Frances Perkins, wrote later in *The Roosevelt I Knew:* "He looked like an invalid who has been allowed to see guests for the first time and the guests had stayed too long."[27]

As he departed for Yalta in early February to determine the destiny of Europe, Roosevelt looked gravely ill. In photographs, Winston Churchill and Joseph Stalin—who look hale and vigorous—appear to be hovering over the thin, drawn president. Roosevelt's famous cape is askew, appearing more like an invalid's blanket. Churchill's doctor, Lord Moran, made these notations in his diary:

> The president looked old and drawn; he had a cape or shawl over his shoulders and appeared shrunken. He sat looking straight ahead with his mouth open as if he were not taking things in. Everyone was shocked by his appearance. . . . To a doctor's eye, the President appears a very sick man. He has all the symptoms of hardening of the arteries of the brain in an advanced stage, so that I give him only a few months to live. But men shut their eyes when they do not want to see, and the Americans here cannot bring themselves to believe that he is finished. His daughter thinks he is not really ill, and his doctor backs her up.[28]

At Yalta his blood pressure was 260/150, a level we now call malignant hypertension.[29]

With such a strong need for and devotion to their president, Americans were ready to believe what they wanted to believe: that Roosevelt was healthy. But severe blood pressure elevation such as he was experiencing can cause chest pain, like the attack he had experienced a few months earlier in Bremerton, as well as congestive heart failure, kidney failure, deteriorating mental function, and stroke. It represents a medical emergency. In that condition, the president undertook an arduous 14,000-mile round trip and spent a week orchestrating the final strategy for victory in Europe. He returned from Yalta exhausted.[30] Historians report he was unnaturally quiet, sometimes querulous. He repeated himself as though he had forgotten what he had just said. As the war raged on, the commander-in-chief involved himself in discussions about a new weapon nearing readiness—the atomic bomb.

During the most critical days of World War II, as Roosevelt made decisions affecting the fate of the world, he was suffering from malignant hypertension, coronary disease, and heart failure.

Gray and wan, he headed to his Georgia sanctuary at the end of March for two weeks of rest. The First Lady remained in Washington, but confided to Senator Alben Barkley of Kentucky that the president had lost his appetite, and had taken to eating simple gruel for sustenance. Barkley, who would die of a heart attack himself while speaking at a mock Democratic convention at Washington and Lee University in 1956, agreed that Roosevelt was getting far too gaunt.

Three reporters were allowed to go to Warm Springs on the condition that they would dispatch no news—including the very fact that the president was in town. Americans believed Roosevelt was still in Washington. Those who saw him during his last days were shocked at how aged he looked. He was down fifteen pounds from his normal range of 184 to 188.[31] His skin was ashen, his lips and fingernails often bluish. Suffering from orthopnea, a telltale sign of congestive heart failure, he had trouble breathing when lying down, and for months had been sleeping with four-inch blocks of wood propping the head of the bed. His hands had a noticeable tremor. The agent at the Warm Springs railroad station, C. A. Pless, accustomed to greeting a smiling, waving man who could never resist the crowd, said later, "The President was the worst-looking man I ever saw who was still alive."[32] Roosevelt barely nodded to people who showed up to greet him. A Secret Service agent told Dr. Bruenn that Roosevelt, who had always gracefully boosted himself from wheelchair to automobile with only slight assistance, felt like dead weight as he simply allowed himself to be lifted. Merriman Smith of United Press remembered how Roosevelt's hand shook as he loaded a Camel into his signature cigarette holder and snapped a kitchen match to light it.[33]

On the morning of April 12, Roosevelt donned a dark gray suit, matching vest, and red tie to pose for a watercolor portrait by Elizabeth Shoumatoff. As the artist painted, he signed papers: a State Department letter, a batch of diplomas for distinguished foreigners receiving the Legion of Merit, a sheaf of postmaster appointments. His usual bold signature was uncharacteristically weak. Lucy Mercer Rutherfurd, his longtime companion, was with him, a secret to be kept from Eleanor. He made a little joke, and Rutherfurd smiled. "The last I remember he was looking into the smiling face of a beautiful woman," said Lizzie McDuffie, the maid.[34]

Roosevelt put a cigarette into his holder and lit it. He raised his left hand to his temple, and then seemed to squeeze his forehead. His hand flopped. As he reached for the back of his neck, he said, "I have a terrific headache." Then he lost consciousness.[35] An excruciating headache is a classic symptom of a brain hemorrhage, a catastrophic form of stroke caused by a ruptured blood vessel in the brain. Dr. Bruenn was summoned and within minutes took his patient's blood pressure. The doctor knew the prognosis was terminal. The numbers, an unsustainable 300/190, went well beyond an indication of danger. They were evidence that the tragedy had already occurred. Bruenn, aided by Roosevelt's valet Prettyman and a mess boy working at the cottage, carried the president to his bedroom, laying him in his maple single bed. Two hours later, at 3:45 p.m. EWT (Eastern War Time), the president was dead. Although no autopsy was performed, the cause of death certainly was a massive stroke.

Embalmers noted that the president's "arteries were so severely clogged with plaque that the pump (serving to inject formaldehyde) strained and stopped"; they were forced to inject the fluid from several different spots.[36] When Dr. McIntire reported Roosevelt's death to the press, he once more insisted that the president's health had been excellent, with no sign of imminent danger. He said that the cerebral hemorrhage "came out of the clear sky." White House press secretary Stephen

Early echoed McIntire's surprise in an official statement claiming that "the President was given a thorough examination by seven or eight physicians, including some of the most eminent in the country, and was pronounced organically sound in every way."[37]

Whether perpetuated through medical ignorance or a deliberate smoke screen, the "splendid deception" around FDR's health continued right up to his death. His unexpected demise undoubtedly fed the public belief that cardiovascular disease strikes quickly and without warning. During his four terms as president, he progressed from garden-variety high blood pressure to malignant hypertension—something almost never seen anymore. He suffered from heart failure and died of a brain hemorrhage, without receiving any effective treatment for his hypertension—because none existed.

Roosevelt died in his prime, within sight of the war victory he helped forge, and the peace he had hoped to underwrite. Eleanor Roosevelt telegraphed her four sons in the service, telling them their father "had slept [sic] away this afternoon. He did his job to the end, as he would want to do."[38] Her first words on hearing the news were "I am more sorry for the people of the country and of the world than I am for us."[39]

The story of Roosevelt's uncontrolled hypertension continues to shock and move listeners. I recently gave a talk at the Callahan Senior Center in Framingham to a crowd that included seven original participants in the Framingham Heart Study. These are people old enough to have firsthand memories of the president, and after I told them about our powerlessness to help Roosevelt as his blood pressure rose, William Walenski approached me.

"In 1943, I was in the Marine Corps at the Opa-Locka, Florida, Naval Air Station," he recalled. He remembers being alerted to serve on a guard detail, being armed with a submachine gun, and boarding a bus to the Pan American Airway Station in Miami for duty. It was daybreak when he saw three planes come in for a landing, two of which took a right flank while one took a left flank, landing near him among the row of ready sentries. It was only when he observed, coming down the gangplank, officers adorned in more top-level brass than he'd ever imagined, followed by a wheelchair, that he realized it was the president of the United States.

"I saw him close up as they wheeled him off the plane in a wheelchair. He was smoking a cigarette. He waved, you know, the typical Roosevelt way—a big wave with his cigarette holder.

"But he was pale. Whiter than white," Walenski said. And all these years later, he still became choked up at the memory. It was only in retrospect, after the discussion at the senior center about FDR's health problems, that Walenski was able to recall the scene with fresh insight. "Before, I had only seen him in pictures, and of course you never saw the wheelchair in pictures. Now as I look back, he did look very tired and sick. You could tell something was wrong."[40]

About two years after that unexpected call to guard duty, when Walenski was far from home, came the moment that he and his generation would never forget. His eyes reddened and filled with tears as he said: "I was in Okinawa in heavy combat on April 12, 1945. I get emotional even talking about it today."

When Roosevelt died doctors had little more than folk wisdom at their disposal to control blood pressure. To look at the picture of cardiovascular ignorance just six decades ago is startling. Heart disease, the most common form of cardiovascular disease, was so ubiquitous that it was considered an inevitable consequence of aging. In 1932, when Roosevelt was elected to his first term as president, a study of medical records in Washington, D.C., showed just how inept physicians were at diagnosing and treating hypertension. That year, only 527 deaths were officially recorded as blood-pressure-related.[41] Modern estimates of the true toll of hypertension at the time put the figure at 140,000 deaths a year.

Life expectancy, which should have soared following the introduction of antibiotics and vaccines, was held back by heart disease and stroke. No one had a clue about the source of the destructive power of these diseases, or what to do about them. Roosevelt became a symbol of the vast uncharted territory of cardiovascular medicine. In the postwar years, his death would serve as a wake-up call, shocking Americans and uniting scientists and politicians behind a massive research effort.

TWO

The Dawn of Peace and Prosperity— and a Deadly Lifestyle

In preparation for the fiftieth anniversary celebration of the Framingham Heart Study, I scanned hundreds of yellowing newspaper clips and old photographs. I was often invited then to speak about the history of the Study, and I wanted to remind myself of that time and place, of just how far we had come in understanding cardiovascular disease. One photograph in my file featured two people whose names I knew well—Walter Sullivan and Victor Galvani. As young men, they served as community leaders who helped and advised the Heart Study. Both remain active in the project to this day. Old pictures provide an especially stark reminder of the passage of time. These men, whom I know as patriarchs, as grandfathers, as community pillars—yes, and as white-haired, respected elderly gentlemen—were, in their prime, the promise of the future. They and their families were representative of a generation on the cusp of revolutionary lifestyle changes. They were typical of men returning from World War II to start families and share in the new prosperity.

In Framingham and across America it was a time of unbridled optimism, a time when a nation, long deprived, could let the good times roll. Fields that once grew potatoes, corn, and soybeans began to sprout duplex, Georgian, Cape Cod, and ranch-style homes, making way for a new generation that would come to be called, collectively, the baby boomers. Bulldozers uprooted thickets of trees, creating stark landscapes that filled practically overnight with row upon row of cookie-cutter houses. Construction techniques that were perfected in the erection of prefabricated military buildings spread across the land.

Houses for the newly discharged veterans went up assembly-line style. Crews of workers raised two-by-four frames and nailed sheetrock to one side, shingles to the other. They moved from one house to the next to create whole new communities of dwellings uniform in everything from bathroom faucets to kitchen cabinet hardware. Little more than the saplings that would eventually break the monotony of the barren yards held the promise of individuality. After the war, the federal government promised 2.7 million new houses by 1948,[1] and loans granted by the Veterans Administration and the Federal Housing Authority made these houses available for little or no money down. The returning GIs needed this benefit. Housing costs had doubled since 1939.[2]

Hundreds of thousands of men and women who had put off college to go to war returned to school, and hundreds of thousands more who had never dreamed of college—some who hadn't even finished high school—earned degrees and entered the middle class. They did it courtesy of the GI Bill of Rights, which provided $500 in yearly tuition and a living allowance of $90 a month for a married man.[3] Walter Sullivan, a slight son of Irish immigrants, and Victor Galvani, the sturdy offspring of Italian immigrants, were two veterans from Framingham who practiced law in Framingham after the war. Where once, as children, they were members of rival ethnic "gangs" whose mischief was kept under control by a beat cop on foot patrol, they grew to be professional men who used their talents and spare time to serve together on a citizens' committee that helped enlist five thousand of their friends and neighbors in the Framingham Heart Study.

This long-term observational Study was aimed at sorting out the reasons behind the epidemic of heart disease and helping many millions beyond the handful of participants directly involved. Eventually, its accomplishments would be listed among cardiology's ten greatest achievements of the twentieth century in an editorial in the *Texas Heart Institute Journal,* in the company of many other remarkable breakthroughs: the electrocardiogram, coronary care units, open-heart surgery, implantable cardiac defibrillators, and coronary angiography.[4] The Heart Study would also be listed in *The Merck Manual*'s Centennial Edition as fourth among the one hundred most significant advances in all of medicine in the twentieth century, along with such accomplishments as the development of antibiotics, the use of mass immunizations, and the discovery of vitamins. The Framingham Heart Study would invent a way to use epidemiology, or the study of diseases in the population, a discipline that had heretofore dealt with the simpler mysteries surrounding infectious diseases like cholera and tuberculosis. Framingham would push the envelope of epidemiology. Its researchers would learn and create ways to observe healthy people over the course of their lives, collect data, and come up with a list of factors contributing to disease. In the process, the study would help create a way of using the science of epidemiology to understand not only heart disease, but many other complex, multifaceted illnesses. "It is truly one of the great studies of this century," said cardiac surgeon Michael DeBakey. "It has set the model."[5]

But none of the researchers or volunteers involved in 1948 knew whether success or failure lay ahead when Sullivan, Galvani, and others asked their neighbors to participate in the fledgling project.

There was much that postwar Americans craved, and much more they didn't know. Pent-up consumer demand resulted in a nationwide buying spree as young families spent their savings on homes, furnishings, children, cars, and a cascading array of time-saving appliances. The modern world was in the process of engineering most of the physical work and exercise out of day-to-day life.

Ironically, they were about to enter what would become for millions a life-threatening prosperity. Their fantasies of the good life, their collective yearnings and imaginings, would take many down a path toward ill health, disability, and even premature death. After the deprivation of the Great Depression and the rationing of World War II, men felt good about being able to provide for their families. And women, having been dismissed from wartime factory work, went home, donned aprons, and rustled up daily high-fat feasts. A glimpse inside one of those postwar households reveals the extent of our misconceptions about what constituted healthy living.

A typical American breakfast was fried eggs with bacon or sausage and a side of toast slathered with butter. Housewives skimmed the cream from the top of milk bottles for coffee. Children washed down their after-school cookies with whole milk. Eggs and bologna, fried in butter, made a tasty lunch. Steak and mashed potatoes were seen as a hearty and healthy meal. Ethnic variations included corned beef and cabbage, chopped liver, and lasagna. Children would fight over who got to eat the crispy fat trimmed from the fried pork chop or steak. Ice cream, once a rare treat, then a reward for being good, soon became a typical, often expected, way to end a summer day. Picnickers ate franks and beans. Every housewife had a crock near the stove to hold bacon drippings, lard, and Crisco to use and reuse in deep-frying everything from potatoes to chicken. Americans believed these foods were good for them. Serving sizes grew from adequate to huge to gargantuan. After dinner, 70 percent of men sat back and lit up a cigarette.[6] Women, who had begun smoking in the 1920s, would start to catch up with men over the next two decades. The whole family got into the habit of sitting around the living room, transfixed by the black-and-white television set. By 1949, Americans were buying 100,000 television sets a week, and by the mid-fifties, two-thirds of families owned one.[7] The first generation of couch potatoes was born.

Americans were getting fat and comfortable. Battalions of veterans and young married couples were heading to the suburbs, expanding the metropolitan landscape outward. Real estate developer William J. Levitt added an eight-inch television set and a washing machine as a deal clincher in his prefabricated Long Island community, Levittown. In a stable community like Framingham, it took most of the first fifty years of the twentieth century for the population to grow from 18,000 to 23,000. In only the first five postwar years, the town would grow by another 5,000 as farmland made way for housing developments, shopping centers, and highways.[8] The new homeowners kicked the frugal habits of their parents and grandparents and filled their homes with laborsaving devices. Power lawn mowers replaced manual rotary models, and soon the mowers were self-propelled. Housewives bought Hoover and Electrolux vacuum cleaners. The chores of daily existence required less and less physical effort even as portions on the plate grew in quantity, calories, and fat content. By the end of the 1950s, four out of five American families owned an automobile, and an interstate highway system was growing and widening to accommodate the vehicles. In the suburbs, officials were literally rolling up the sidewalks, providing no safe or convenient way to get around without an automobile. Some Americans played golf or tennis, but chores required less and less physical exertion. Even walking to the bus stop or the trolley line was becoming an obsolete exercise. "We walked to church, two miles, the whole family, every Sunday morning," says Evelyn Langley, now eighty-seven, of Framingham. "Then we became a little more affluent. We got a car and we would ride down to church." It seemed that everything was conspiring to make people more sedentary.

At first, families had just one car, and few women worked outside the home. So food came to them, bells or music echoing down the block to announce the arrival of fresh treats. "The Happy Homes Bakery truck would come by," says Karen LaChance of Framingham. "Everything came through the neighborhood. Ice cream trucks. Vegetable trucks." LaChance, who refers to her ethnic roots as "shanty Irish," is now a professional woman who works out regularly and eats sensibly. But she wasn't always so careful. She remembers her adolescence as a time of packing on the pounds. After-school snacks were cupcakes and goodies from the bakery truck, accompanied by what every schoolchild considered to be nature's nearly perfect food: whole milk. "Every year, I gained another ten pounds. I got to be 190 pounds by the time I was a freshman in high school. I was a chunker," she recalls.

Supermarkets were opening on main streets, pushing aside motorized peddlers as well as mom-and-pop grocery stores. Corporate food producers were making life even easier with products like Velveeta and Spam. Potato chip production rose from 320 million pounds in 1950 to 532 million pounds in 1960. During the 1950s a uniquely American problem arose from the country's recent affluence: how to dispose of up to 100 million tons of surplus farm produce. Food products with little nutritional value began appearing on grocery shelves. Americans were about to prove that obesity was simply a different form of malnutrition.

Even cigarettes, which were a rare indulgence at the turn of the twentieth century, when smokers rolled their own and the safety match was not yet invented, became ubiquitous. Men who had been to war smoked because cigarettes were part of their K-rations. Tens of thousands of veterans came home addicted to nicotine. Movies, television, and advertising encouraged the habit, and helped spread it to millions more men and, increasingly, women. Viewers watched Edward R. Murrow bring integrity to television as he spoke through clouds of cigarette smoke during his program *See It Now.* Murrow chain-smoked through interviews in an even more popular program, *Person to Person*. Lucy Ricardo also nervously lit up as she and Ethel Mertz concocted schemes so crazy they just might have worked. *I Love Lucy* was so popular that the Chicago department store Marshall Field & Co. switched its weekly clearance sale from Monday, when Lucy was on, to Thursday.[9] "TV, a twelve-inch model, was my babysitter," says Faye DeSaulnier, fifty-one, of Framingham, a second-generation volunteer for the Heart Study. During commercial breaks, DeSaulnier, as a child, would watch a dancing Old Gold cigarette pack or hear a "Call for Philip Morris." "Winston tastes good like a cigarette should" violated the grammatically correct "as a cigarette should" and drove English teachers crazy. L.S.M.F.T. meant "Lucky Strike means fine tobacco," and more doctors smoked Camels than any other cigarette. People were happy to indulge in the satisfying taste of fat, to enjoy a cigarette, and to take it easy in front of the TV.

The good times were rolling, and no one wanted them to stop. Americans had homes, jobs, and peace. They were prepared to indulge themselves in every luxury and excess they had coveted for decades. They worked hard, but the very availability of work was a privilege, and America was a nation of people who felt they deserved a little reward. And if a little felt good, a lot felt even better. Satisfaction became overindulgence, and overindulgence spilled into conspicuous consumption.

For health care, most had insurance they called "hospitalization," because that was pretty much the extent of services it covered. Families took their children to pediatricians for smallpox vaccinations, and soon would do the same for the polio vaccine, but adults went to the doctor only when they felt sick. The notion of an annual physical for a healthy adult was unheard of. Preventive medicine, strongly linked to the discovery of cardiovascular-disease risk factors by the Framingham Heart Study, was still on the horizon.

At a time when excess was finally possible, a small cadre of scientists was getting ready to preach moderation. They worried about the hidden dangers of all this freewheeling eating, smoking, labor saving, sitting, and relaxing. Their concerns were born of prewar animal studies, of autopsies of young soldiers with clogged arteries, and of clinical observations. The life insurance industry provided clues about premature death from its perspective. After all, the businesses writing out large checks for deaths that came too soon had a stake in seeing subscribers pay their premiums for many more years. They were on the front lines of understanding why so many died too soon. In a medical journal report, Louis I. Dublin wrote: "Because of our prosperity and abundance, a large number of our people are literally eating themselves to death."[10] Some medical researchers had their educated suspicions about possible contributors to the emerging epidemic of heart disease.

Physicians across the country had endured their own frustrations and deprivations during the Depression and war years. Immediately following the war, researchers had little more than theories, and general practitioners had nothing to offer patients with heart disease.

One of the few "safe" possibilities for treatment was a strict diet to reduce weight and lower blood pressure. In 1939, Walter Kempner came up with a protein-free diet of rice, fruit, and vitamin tablets at a time when prominent nutritionists were enamored of protein and wary of carbohydrates. Kempner called it the Duke diet, and it put Durham, North Carolina, on the map for those unfamiliar with the university. The diet, and the institution, drew celebrities, including Lorne Greene of *Bonanza,* Dave Garroway of the *Today* show, cartoonist Al Capp, actress Shelley Winters, and, arguably one of the contributors to the problem, Colonel Harlan Sanders of Kentucky Fried Chicken. Frank Sinatra sent his overweight mother. In the lay press and the Durham community, the dieters became known as Ricers, and the popular press was more interested in the treatment than were its medical counterparts. *Esquire* put news of the diet on its cover, and *Harper's Bazaar,* taking note of the wealthy clientele, mentioned it in a column called "The Idle Rich." Betty Hughes, wife of New Jersey governor Richard Hughes, wrote of her experience as a Ricer for the *Ladies' Home Journal* in 1968. "By the end of the third day, the rice, which is washed several times before and after cooking to remove the last traces of starch, began to taste like damp Kleenex," she wrote. The diet and its associated expenses, including lodging, cost $1,000 a week.[11] Clearly, it was not the universal answer the country needed.

It turned out that the diet, which was virtually salt-free, lowered blood pressure, but was unpalatably bland for most patients. Even Kempner, a flamboyant recluse who drove a Lincoln convertible and nearly always wore a trademark blue blazer and white shoes, admitted it was uninteresting fare.

With little evidence available, experienced physicians across the board advised against treatment of hypertension. Certainly in 1948, and even as late as the mid-1960s, textbooks were telling young doctors that there was no point in lowering blood pressure if there were no symptoms, and that no one over the age of seventy should be treated for high blood pressure. Giants in the field, including Franz J. Ingelfinger and Arnold S. Relman, who each served as editor of the prestigious *New England Journal of Medicine,* edited a book, *Controversy in Internal Medicine.*[12] It was 1966 when physicians argued in a chapter on high blood pressure that treatment of hypertension did no good. They instead urged science to find the underlying cause of hypertension, and then treatment would be obvious. It's a good thing science didn't wait. As of this writing, the cause of most cases is still unknown, but 25 million Americans now receive beneficial treatment.

As leading scientists were gearing up to unravel the mysteries of heart disease, the general practitioners of the day were powerless to treat what had become the leading cause of death. Conventional wisdom held that hypertension and its fatal or debilitating consequences—heart attack, congestive heart failure, and stroke—were the unalterable results of aging. In those days, severe hypertension would accelerate and escalate to levels few American physicians see today. As the disease ran its unstoppable course, patients' eyes would hemorrhage, their kidneys would fail, and at last their hearts would stop. "It was very sad in those days. We'd put them in a dark room, give them a sedative, and wait for them to die," recalls Edward Freis, the researcher who proved that lowering blood pressure saved lives.[13]

The vast majority of physicians had far less expertise and far fewer clues than did the small but growing cadre of researchers and heart experts. Physicians treated patients largely on the basis of anecdote or observation within their own practices. Some cautioned against eggs. Some advised exercise. Others thought exercise was dangerous. Young men who were athletic enough to earn a letter in a sport during their college years were found to be more prone to heart attacks later in life. But was that due to exercise itself, or to later obesity and sedentary lifestyle among former athletes? No one knew. Early reports from

France led some to suggest drinking moderate amounts of wine. But the numerous theories were unproven, and when a patient's heart deteriorated, "We'd take him to the hospital, bring in a priest, and start praying that he gets well," says Menard Gertler, a cardiologist from Montreal who came to Boston to work at Massachusetts General Hospital shortly after World War II. Usually, the patient didn't get well.[14] The death certificate would be as confused as the treatment decisions. "When we started, we were getting death certificates saying that patients had died of acute indigestion," says William B. Kannel, who joined the Framingham Heart Study in 1950. Or "apoplexy." Or just plain old "collapse," when in fact the cause of death was heart disease.[15]

How could these researchers begin to tell a generation who never before had it so good that much of what they liked was bad for them? Worse than bad. How could they spread the word that the very indulgences that made so many feel rewarded were threatening to kill them? It would take the better part of a quarter of a century to come up with the proof. And that was the relatively easy part. It would take nearly twice as long to begin to persuade large segments of the population to abandon their comfortable but increasingly sedentary lifestyle, to cut back on their rich, abundant foods, to permanently put out their cigarettes, and to go to the doctor for routine checkups even when they felt perfectly fine. The battle for the hearts and minds of Americans continues, and as the twenty-first century commences, we have largely lost the war against obesity.

The foundation for the research was laid with the creation of the National Institutes of Health, which would quickly become the world's foremost medical research establishment. Some ninety-two acres of an estate called Tree Tops, in Bethesda, Maryland—a 1935 gift of Mr. and Mrs. Luke I. Wilson to the federal government—stood ready to house the newly named National Institute (singular) of Health. President Roosevelt dedicated the buildings and the grounds on October 31, 1940. Today the campus of the National Institutes of Health has grown, through additional land purchases, to 306 acres.[16] During the war, the NIH focused on war-related problems. Why, for example, were 43 percent of young American men deemed unfit for general military service—28 percent unfit for any military service? Any hope for new and expensive research into non-war-related health problems, like cardiovascular disease, was out of the question.

With the end of the war, scientists sought the freedom and financial support to pursue their research, and high on the list was tracking down what was responsible for cardiovascular disease, the first noninfectious epidemic to sweep the country. Costly ideas were on the table because of postwar prosperity. Vannevar Bush, director of the Office of Scientific Research and Development, was a powerful voice in pleading with President Harry Truman to make medical research a national priority. "Notwithstanding great progress in prolonging the span of life and in relief of suffering, much illness remains for which adequate means of prevention and care are not yet known," he argued. "While additional physicians, hospitals, and health programs are needed, their full usefulness cannot be attained unless we enlarge our knowledge of the human organism and the nature of disease. Any extension of medical facilities must be accompanied by an expanded program of medical training and research."

Bush urged the federal government to support research, where unexpected discoveries could lead to understanding, prevention, and cure. He envisioned a system where researchers would be free to follow their curiosity and imagination. "The history of medical science teaches clearly the supreme importance of affording the prepared mind complete freedom for the exercise of initiative," he wrote.[17] Bush knew the value of gathering brilliant minds and focusing them on an urgent task. During the war, he had supervised the Manhattan Project, which led to the atomic bomb. With that background, he was perfect for the task of solving the country's most urgent peacetime problem: slowing the epidemic of heart disease and applying science to understand how to prevent it.

He set the tone for the establishment of a national research institute built around the optimistic scientific assumption that general, basic knowledge would ultimately lead the way to specific answers, treatments, and cures. In the twenty years before Roosevelt's death, childhood mortality rates had dropped by 87 percent, shifting the national research focus to the killer disease that was attacking middle-aged adults. Massive unemployment and homelessness and the war itself were history. Now, the largest single threat to the nation's health was cardiovascular disease. Researchers in the U.S. Public Health Service, the parent agency of the NIH, were full of ideas. The country's medical researchers were more than ready to take on heart disease. Congress passed the National Heart Act in 1948, creating the National Heart Institute (now the National Heart, Lung, and Blood Institute). When Roosevelt died in 1945, only 374 physicians in the United States specialized in cardiology.[18] Four years later, in 1949, the American College of Cardiology was established. Today its membership numbers 22,000 heart specialists.[19]

The field of heart disease research was unexplored territory, a blank slate where young scientists were free to create and discover and to make their marks.

These investigators had only the faintest scent of the killer. Long before Roosevelt's death, scientists had been trying to understand heart disease. Pathology research had been going on for two centuries and animal studies went back seventy-five years. Nearly a half century earlier, the electrocardiograph was introduced, giving researchers a way to measure cardiac damage in a living person. None of it, though, answered the question of what caused heart disease.

A handful of bold thinkers had ideas that went against conventional beliefs. Ed Freis believed treating high blood pressure could work, without doing harm. Kilmer McCully argued that homocysteine had a primary relationship to the damage in arteries that led to heart disease. Ancel Keys and Jeremiah Stamler felt that the food people ate, especially the kinds of fats they ingested, was related to their risk of heart disease. Michael Brown and Joseph Goldstein thought that by studying cholesterol receptors on cells, they could figure out the process by which LDL, or bad cholesterol, became overabundant and built up as life-threatening plaque within artery walls.

Eventually, some of these scientists would receive the Nobel Prize. In keeping with a centuries-long tradition of punishment for scientific heresy, some would suffer ostracism and ridicule from their peers. But the majority would soldier on in a field that denies instant gratification, waiting years for proof and reward of their efforts, and beset by political battles. Despite numerous setbacks, progress continued.

Before the threads of past research could come together, medical science had to understand what was different about those who developed heart disease, compared with those who didn't. Congress, no longer bogged down by the war, was ready to fund the effort. In 1948, it passed a bill to grant $500,000 to the NIH to uncover the root causes of heart disease—the seed money for the Framingham Heart Study. Another bill of similar size allocated $492,000 to fight the Long Island potato bug.[20] Today, nearly all of the potato farms of Long Island are gone. Old farmlands have been turned into lucrative vineyards on the north fork of the eastern end of the island. The south fork has become a summer refuge for the well-heeled. Potato bugs are no longer a priority.

But heart disease is. Understanding the root causes would mean sorting through theories and painstakingly proving which of them was correct. That was the Framingham Heart Study's undertaking. Just over a decade after the Study's start, the early director, Thomas R. Dawber, along with his successor William Kannel, had coined the term "risk factor." By the end of the twentieth century the term was part of the American lexicon. The answers so far, more than half a century into the unraveling mystery, seem patently obvious. They are accessible to everyone, rich and poor. Millions of lives have been prolonged because of the findings. Yet countless millions of those at risk for heart disease steadfastly refuse to follow the now well-established advice about smoking cessation, diet, exercise, and maintaining desirable levels of blood pressure and cholesterol. But we all know enough at least to feel guilty about what we're not doing. The future may help pinpoint just who needs to feel guilty, and who can safely disregard prudent medical advice. After all, as Nobel laureate Michael Brown points out, Winston Churchill, who smoked, drank heavily, ate heartily, and never exercised, lived to ninety-one. Jim Fixx, who practically invented jogging for fitness, keeled over on a run, dead of a heart attack at fifty-two.

More than half a century after FDR's death, the basics are known. "I actually feel like a criminal if I eat the skin on my chicken," says Karen LaChance. Even as restaurants specializing in steak once again proliferate, those ordering sirloins and T-bones know that fish is a healthier option. Those eating designer ice cream with up to twenty grams of fat per serving are aware that it's sinfully good. Prime rib, doughnuts, and onion rings come with a side order of guilt. A U.S. Department of Agriculture slogan about fruits and vegetables, "Strive for five," has caught on enough so that consumers know when they're falling short. Exercise has become a multibillion-dollar industry, even as scientists prove that frequent spirited walks will do the trick for free. The food industry is beginning to cooperate. Consumers can choose full-fat milk, or taper down to 2 percent, to 1 percent, and to fat-free skim milk, without losing any important nutrients. The FDA now requires food labeling that helps consumers calculate daily fat intake.

Smoking, still troublesome in the number of youngsters starting the habit, has been on a downward slide. And the use of widely available drugs to control blood pressure and cholesterol levels is steadily rising. People are getting in the habit of seeing a doctor for a checkup routinely.

A few years ago, I was on National Public Radio's *Talk of the Nation Science Friday,* with host Ira Flatow. The five-decades-long contributions of the study participants were on my mind. As coincidence would have it, a 1949 photograph of the youthful Walter Sullivan was on top of my pile of papers when a caller from Salt Lake City came on the air.

> Hi. This is Dick Sullivan. I thought you might like to hear from a participant in the heart study. I'm an offspring member. My dad is Walter Sullivan. He was the chairman of the citizens' committee in '48. My brother is still in it, my sister is in it. I have a little anecdotal story. My dad had open heart surgery in 1993 when he was seventy-nine. And as they treated him, they—the doctors at Framingham Union Hospital— said that he had the body of a forty-year-old guy. Because of that, he responded to treatment so well. And I guess the moral of the story is he learned a lot of good things in the Heart Study—adjusted lifestyle, quit smoking about 1960—that helped him out. I can remember a quantum change for my dad. He used to cut the steak for us as a family and eat the fat.

At most family dinner tables now, the fat gets scraped into the garbage can. A quantum change was in store for everyone. At the midpoint of the century, however, it had yet to unfold.

THREE

Gathering Evidence, Building on Clues

I could go into virtually any town in the world, measure a few simple things on people's bodies, and identify most of those in that town who are going to get a heart attack or stroke," says William Castelli. "No. Better than that, I could *lower* those bad numbers and *prevent* many of those heart attacks and strokes."[1] Castelli, director of the Framingham Heart Study from 1979 to 1994, made that statement more than fifty years after the best medical minds could do nothing to save President Roosevelt.

How did we get, in half a century, from medical powerlessness to the development of an arsenal of techniques that, if used to full effect, can prevent and control heart disease? Today, presidential blood pressure numbers like FDR's would send the country's leading doctors racing down hallways, ordering helicopters, shouting commands into cell phones, and whisking the nation's leader into the cardiac care unit of Bethesda Naval Hospital. The medical armamentarium includes hundreds of new drugs, sophisticated coronary care units, implantable defibrillators, electrocardiographic monitoring, coronary artery bypass surgery, coronary angioplasty, and the implantation of stents to improve the chances of keeping the artery open longer. And the state of the art keeps evolving. Patients get sent home from the coronary care unit with instructions to take aspirin daily as well as prescriptions for cholesterol-lowering drugs and beta-blockers and ACE (angiotensin converting enzyme) inhibitors to help prevent second heart attacks and heart failure. And the modern prevention-oriented medical strategy offers a range of options to ward off cardiovascular disease by lowering blood pressure and cholesterol levels. Even the common aspirin has been elevated, thanks to research, to the level of a powerful heart disease preventive agent.

In the late 1940s, at a time when researchers knew little about the causes of the disease, they at least understood that the factors behind it would be numerous, varied, and complex. Many of the leading cardiologists and scientists in the country made educated guesses about which avenues of study to pursue. This challenge was not like locating the origin of a cholera outbreak, where one physician, John Snow, sticking pins into a map of London where victims of the epidemic lived, traced the root cause to a contaminated water pump. Once he understood the source, the prevention of future outbreaks was as simple as removing a pump handle to stop the flow of disease. Those seeking the sources of heart disease were facing dozens of contributing causes and countless possible preventive techniques and treatments.

By this time, there was a steady stream of research on which to build. For the few who focused on cardiovascular research, the hints from papers published over many decades pointed in the same direction. A Russian laboratory study of rabbits showed that atherosclerosis caused cardiovascular damage. Other laboratory bench research and animal study clues gave weight to autopsy results indicating the central role of atherosclerotic plaque. Studies of populations, noting disparate rates of heart disease around the globe, made a lot of researchers wonder if what people were eating, as well as other lifestyle factors, affected their hearts.

As early as the eighteenth century, scientists examined cadaver aortas and saw the buildup of plaque. They used words like "cheesy," "soft," and "mushy" to describe the accumulation in earlier stages and "hard," "firm," and "calcified" to characterize its later look and feel. Indeed, *athēra* means "mushy" in Greek and *sklērōsis* means "hardening," an oxymoron within a single word. These early scientists, however, didn't know anything about the causes of atherosclerosis. But their suspicions and early observations directed others toward a path of discovery.

An English physician named William Heberden built on eighteenth-century descriptions of hardening of the coronary arteries. He described in detail the symptoms of chest pain related to this process, though he laid no claim to understanding what caused it. In 1768, more than a century and a half after the first permanent English settlement was founded at Jamestown, Virginia, Heberden first used the term "angina pectoris," and his description of the characteristic discomfort due to coronary disease endures. "They who are afflicted with it are seized while they are walking (more especially if it be uphill, and soon after eating), with a painful and most disagreeable sensation in the breast, which seems as if it would extinguish life."[2] Another Englishman, John Fothergill, read Heberden's paper and proposed further that extreme emotions like anger and anxiety could bring on the pain as well.

Across the ocean, American colonists were taking up their muskets against King George III's army. They could not have paid much attention to what English physicians were describing as a disease that came on during exertion or high emotion, struck the heart, caused severe discomfort, and often resulted in quick death. In America and the rest of the world, humankind was absorbed with survival. And the medical challenge laid down by Heberden, however provocative, was ignored for nearly one hundred years.

In the mid-nineteenth century Rudolf Virchow opened the doors of cellular pathology. His discoveries spanned dozens of diseases. About the time American pioneers were acting on the words of Horace Greeley, "Go West . . . and grow up with the country," Virchow published, in 1859, his book *Cell Pathology*. It became the foundation for microscopic study of disease. Because of him, medical students today study pathology. His work allowed scientists examining the plaque in arteries to discover that lipids, more specifically cholesterol, and even more specifically, cholesterol esters, made up the core of the atherosclerotic plaque that blocked the flow of blood to the heart or brain.

Virchow and his protégés were working with rabbits, injecting them with adrenaline to damage cells in the arteries so they could create and observe atherosclerosis. "They didn't do very much with that," says William Castelli, something of a cholesterol historian. "But they had befriended the rabbits, and they started feeding them meat scraps. And those pampered rabbits were the ones that keeled over from atherosclerosis." In the mid-nineteenth century, Americans, when they could afford meat, were eating the relatively lean cuts of free-roaming livestock. The more adventurous among them, as they explored and conquered the frontier, were sampling the even more well-exercised meat of buffalo. It is only during the twentieth century that our marbled, tender meat began to come from sedentary cattle.

At the beginning of the century, the leading causes of death were infectious diseases: pneumonia and influenza, tuberculosis, diarrhea, and intestinal diseases. "Diseases of the heart" were listed as the fourth-leading cause in 1900.[3] The annual death rate from diseases of the heart was 137 per 100,000—fewer than the 260 per 100,000 when the Framingham Heart Study began; far fewer than the peak of the epidemic in about 1970 when 340 per 100,000 were dying of heart disease; and fewer than the 200 per 100,000 lives currently claimed each year.[4] But the numbers are not comparable. Heart disease may well have been a bigger problem than anyone knew. The epidemic, if there was one at the beginning of the twentieth century, was hidden by a life expectancy of only forty-seven years, not giving heart disease enough time to take hold. It was also concealed by misdiagnoses and misattributions as to cause of death. Without accurate diagnosis in living people and with inaccurate causes listed on death certificates, cardiovascular disease was getting away with murder.

By the time of the first meeting of the International Congress of Pathology in 1904, scientists understood from the work of pathologists and organic chemists that atherosclerosis meant a buildup of plaque that was rich in cholesterol esters. America was transforming itself from a rural republic to an urban nation, when scientists were beginning to recognize a pattern; those who succumbed to heart disease commonly had symptoms that predated their disease, often by years. But had there been a radar screen of worries in American minds at the time, heart disease would hardly have raised a blip.

When heart disease did strike, physicians could do nothing, but some scientists had a hint of causes. Paul Dudley White spoke in 1951 of looking over medical records from 1911 to 1913. "I have come across cases of heart disease that were diagnosed simply as hypertrophy and dilatation of the heart or valvular disease, without any statement of cause."[5] And without an appreciation of cause, treatments were primitive. "Our frequent misuses of digitalis would be amazing to young physicians of today. We had no good diuretics. We did not limit salt, though we did make patients uncomfortable from thirst by limiting fluids. Such were the good old days!"[6]

In the early years of the twentieth century, even ordinary physicians, largely within the developing specialty of internal medicine, were noticing a link between high levels of blood cholesterol and five other conditions: hypothyroidism, diabetes, bile duct obstruction, kidney failure, and a rare disorder of extremely high cholesterol levels called familial hypercholesterolemia. Patients with these illnesses had one thing in common: they were at risk for premature atherosclerosis. These observations posed a major question. Does the mushy deposit that accumulates in the arteries come from the circulating blood? It became an area for investigation.

Then came a finding that would one day be labeled a landmark, but in its time didn't even stir a ripple of interest. In the December 7, 1912, issue of the *Journal of the American Medical Association,* James Herrick compared the symptoms of living patients with those who had died and, upon autopsy, were found to have blocked coronary arteries. His conclusions went against the nearly unanimous thinking of the day that a heart attack meant certain death, usually within minutes. The *JAMA* article was five and a half pages of understatement and began with scant evidence: when the coronary arteries of animals were blocked with clamps or artificial emboli, rabbits and dogs quickly died in most experiments. But in one study the rabbits survived the blockage, and in another some dogs with similarly blocked arteries lived. Cases were documented in which pathologists found that some patients were autopsied with complete obstruction of one coronary artery, and yet died of other causes with no apparent cardiac symptoms. This was the evidence Herrick adduced as he proposed a radical theory. His modestly written article continues to be credited as the first modern description of myocardial infarction, or heart attack, an episode in which insufficient blood supply causes death to a part of the heart muscle.[7]

Herrick argued that a disease believed to be always fatal could be survived and that perhaps large numbers of heart attack victims were still walking around. They might survive, he suggested, for hours, days, months, or even years.

The insight came to him in the person of a fifty-five-year-old banker. The man had just returned from the theater in Chicago, and capped the evening with a sandwich and a bottle of beer. Herrick saw him on January 10, 1910, when he was admitted to the hospital with nausea and chest pain. After examining the patient and reviewing his history, Herrick suspected the diagnosis of myocardial infarction. He and another physician, John Murphy, spent the next two days watching the patient, napping in side-by-side twin beds and discussing, as they waited for respites of sleep, what the matter could be. Conventional thinking of

the time said that if he had had an obstruction of a coronary artery causing myocardial infarction—death of heart muscle—he would have died
within minutes.

When the patient rolled vigorously from side to side, got up to use
the toilet, and remained alive, doctors were stunned. They expected
death at any second, and Murphy called the man's son in New York.
Herrick wrote in his recollections: "I shall never forget Dr. Murphy's
emphatic way, as, in his rasping voice, he shouted (the long distance connection with New York was not as perfect in those days as now): 'What's
the matter with you? Get busy, come right away or you'll be too late.' "

The man lived for fifty-two hours after the onset of pain. Herrick
requested an autopsy, and asked the pathologist to look for a clot in the
coronary artery. And that's what was found. "Soon after this, I realized
more clearly that cases of this type were not rare and that the whole subject was worth more extended observation," he wrote. "The more I read
on the subject and pondered over it, the clearer it became to me that
sudden death was by no means the inevitable sequel to the accident."[8]

Herrick's chief tool in drawing his conclusion that human beings
could survive a heart attack was simple bedside observation. Soon,
American hospitals would add electrocardiographic evidence to such
observations. Characteristic changes in the electrocardiographic tracing
of the electrical activity of the heart are diagnostic of myocardial infarction. Most would recognize them as the waves and blips seen in close-
ups on television hospital dramas. But in a heart attack, some segments
rise dramatically, followed in minutes or hours by unique "Q" waves.
There is often some loss of voltage, reflecting death of part of the heart
muscle. But Herrick did not yet have that tool when he and his
colleague, in 1910, observed the slender, active, middle-aged banker
who, after being seized with terrible chest pain, confounded medical
wisdom by remaining alive. All they had were pulse rates, the faint
tones of heartbeats heard through a stethoscope, thermometers, urine
samples—and close observation of the man's condition.

When Herrick published his conclusions, the medical establishment
believed that heart attacks were undiagnosable and untreatable. He cau-
tiously but unapologetically claimed otherwise. His discovery was met
with a deafening silence. Medicine was still operating out of a black bag,
and few were anxious to change the contents. Thirty-five years later,
Herrick submitted "An Intimate Account of My Early Experience with

Coronary Thrombosis" to the *American Heart Journal* at the urging of colleagues who wanted him to create a written record of his discovery that coronary thrombosis—even without life-threatening symptoms— could be diagnosed in a living patient.

> It seemed strange to me at the time, it seems strange to me now, that when, in 1912, I read before the Association of American Physicians a paper that seemed to me to contain an important announcement, it fell like a dud. No one . . . discussed it or even asked a question. I must have been keyed up to a high pitch, for I recall my eagerness to have the article published promptly; I feared someone else might jump into print ahead of me. My anxiety about priority was groundless. Even after its publication in *JAMA,* it aroused no more comment than it did when it had been read six months before.

He said that he was "sunk in disappointment and despair."[9]

But Herrick rallied enough to take his findings on the road, giving lectures and consultations on the medical society circuit, talking endlessly about the type of symptoms associated with coronary disease. He called it his "missionary work." And the reaction of mainstream physicians, he wrote, was this: "A few listened attentively, more, incredulously, the majority, indifferently."

It took a few more years and one more study presented in 1918 and published in *JAMA* in 1919 to catch the attention of the medical community.[10] For that paper, Herrick had the benefit of electrocardiographic tracings to diagnose myocardial infarction. The electrocardiogram took his earlier conclusion beyond conjecture and into the arena of scientific proof. Finally, cardiac specialists took note.

And with that, Herrick gave doctors their first tool and a solid reason to diagnose heart disease *before* the patients died. Thus, scientists were able to contemplate the possibility of reducing the toll of heart disease through treatment. And the more visionary of researchers, chief among them Paul Dudley White, could realistically dream of a day when the root causes might be understood and, better yet, treated to prevent heart attacks. It would be seen as an epidemic, no longer invisible, and the scientific quest to treat and prevent it could begin.

So ended 150 years of medical helplessness in the face of heart disease; an era that might well be considered the dark ages of diagnostic, preventive, and therapeutic cardiology. Between Heberden's description and naming of the chest pain associated with blocked arteries—angina pectoris—and Herrick's characterization of acute thrombosis leading to a heart attack, the two researchers covered both ends of the cardiovascular disease spectrum. But a century and a half had passed without a scientifically backed word of advice to offer a suffering patient.

In his retrospective, Herrick wrote of the terrible frustration of being ignored. "I often asked Dr. Smith and my interns, who shared my enthusiasm, whether I was becoming 'nutty.' They thought not."[11] He was not the first, nor would he be the last, heart disease researcher to go against the mainstream, and end up questioning whether it was he, not the majority, that might be off base.

Yet throughout, his voice remained clear, his evidence strong. Perhaps what prevented physicians from paying attention for so long was the grim truth that even if they could diagnose a nonfatal heart attack, there was nothing they could do to change the direction of a patient's clinical course. Herrick's work was intellectually challenging, but years would pass before his insights moved toward anything approaching practical application. Before his discovery, most victims of coronary disease died undiagnosed or misdiagnosed. When his work was finally noticed and accepted, about a decade later, the statistics on causes of death changed forever. The startling news of a heart disease epidemic would emerge.

The electrocardiograph was a technology that helped prove Herrick right and set the wheels in motion for detecting an epidemic. Thanks to Benjamin Franklin's 1752 kite and key experiment, the science of electrophysiology arose. The heart had something in common with lightning. It had electrical activity. Willem Einthoven used that knowledge to develop a precursor to the modern electrocardiograph machine in 1902, a feat for which he was awarded the Nobel Prize in 1924. His machine took up two rooms, weighed eight hundred pounds, and required five technicians to operate.[12] Today's electrocardiograph is a compact, computerized machine that can be battery-operated and small enough to be hand-carried. Even tinier versions can be hooked onto a belt and worn for twenty-four hours or more to continuously record the electrical discharges of the heart, an innovation that helped pinpoint Vice President Dick Cheney's need for an implantable defibrillator.

When White was in Europe in the early part of the twentieth century, he saw the electrocardiograph machine for the first time. His mentor Sir Thomas Lewis taught White about the importance of the electrocardiogram in diagnosing heart disease. This opened his eyes to the possibilities of preventive cardiology. He persuaded reluctant administrators at Massachusetts General Hospital in 1914 to give him some space.[13] Space is always at a premium at teaching hospitals, and Mass General's best offer was to grant him the use of a closet. White, just twenty-eight at the time, took it. And within his cramped, inelegant quarters, he became one of the first physicians in the United States to use the electrocardiograph for clinical purposes. He would become America's premier cardiologist, eventually treating Dwight D. Eisenhower's heart attack.

But White would also become one of the nation's strongest and most powerful proponents of getting at the causes of heart disease in order to prevent it—an offshoot of his second transatlantic encounter. In Scotland he met Sir James Mackenzie, who, in his practice, tried to conduct an epidemiologic study of cardiovascular disease. The task was overwhelming, but White never stopped thinking about ways that such a study could be accomplished. In 1940 he suggested making a comparison of one thousand Vermont farmers who were working hard physically with one thousand New York bankers who were not.[14] His proposal, interrupted by World War II, was never carried out. Ultimately, he concluded that a practice-based study, such as Mackenzie's, would never provide a broad enough base of sick and well participants. But exposure to the work of the Scottish physician undoubtedly helped him to realize that population-based research was needed. The seed of Framingham was planted in his mind.

When White brought electrocardiographic technology across the Atlantic, it united the work of three visionaries: Einthoven, White, and Herrick. Einthoven invented a miraculous diagnostic device; White brought it to Boston and began using it to diagnose heart disease; Herrick employed the device to prove his theory that human beings could survive a heart attack. White picked up the clues from Herrick's research and added his own vision. He became a proponent of long-term epidemiology as a means to ferret out the causes of heart disease. Together, White and Herrick founded the American Heart Association in 1924.

A year after Herrick's landmark study, a Russian scientist named Nicolai Anitschkow fed rabbits purified cholesterol, which is alien to their preferred diet of vegetables. The animals accumulated deposits of lipids and cholesterol in their liver, spleen, arteries, and other tissue, something that never happens in their natural state. But for thirty years, his findings were considered a curiosity, as was his interpretation, that deposits of cholesterol on artery walls were the main culprits in atherosclerosis. Far from convincing physicians that his research had anything to do with people, he seemed to succeed only in making doctors believe that high cholesterol, an unnatural ingredient in most animal diets, promoted a disease in rabbits. But he remained convinced of his theory. When Donald Fredrickson, former director of the National Heart Institute, met him in the Soviet Union in 1963, Anitschkow was still handing out rabbit aortas to visitors to illustrate his work. He thought cholesterol was the entire heart health answer. In fact, Fredrickson related in a 1993 article in *Circulation* that when the Russian scientist was asked about his constant cigarette smoking, he replied, *"Cholesterin ist Alles"* (Cholesterol is everything).[15]

Ancel Keys, convinced that answers to human cardiovascular disease would never be found by simply watching laboratory rats, began studying the diets of Minnesota businessmen in 1947. He believed the American way of eating was pathological. On a visit to Italy, he saw that the only patients who suffered cardiac problems were "rich men in private hospitals."[16] In Finland, where heart disease rates were high, he watched farmers butter their cheese. His work led him to establish, in 1940, a new field that he called physiological hygiene. It combined physiology, nutrition, epidemiology, and prevention. By the mid-1950s he would be examining the dietary habits of citizens in a project known as the Seven Countries Study. He would eventually log more than half a million miles across three continents. He was building on the observations of earlier researchers. A report from Holland told of formerly vegetarian sailors from Java who signed onto a Dutch sailing vessel, altered their diet to include meat and dairy products, and developed atherosclerosis, uncharacteristic in their native land. A study from Japan, where the diet consisted of little meat and a lot of fish, rice, and vegetables, showed that among its citizens heart disease was rare. Keys spent his career measuring the skin folds of Italians, studying the metabolism of Finns, analyzing the contents of exotic African dishes, and monitoring the diets of his fellow Minnesotans, proclaiming the high-saturated-fat American diet as the villain.

As important as these animal studies and human and autopsy observations were, researchers needed a basic understanding of what was going on at the molecular level. By 1944, Konrad Bloch was beginning to investigate cholesterol molecules, work for which he, along with German biochemist Feodor Lynen, won the 1964 Nobel Prize. He figured out how cholesterol is synthesized in the body from acetate, work that allowed later researchers eventually to develop the statin medications that could lower unhealthy levels of blood cholesterol.

Oddly enough, the field of neurosurgery laid the groundwork for the understanding of the regulation of blood pressure. The majority of physicians still saw rising blood pressure as a necessary physiological response by an aging body in order to force blood through hardened arteries and toward organs. It was thought to be a healthy thing, every bit as good as basking in the rays of the life-giving sun. Leading physicians believed that it was dangerous and irresponsible to lower high blood pressure. That position grew out of scientific dogma from the nineteenth century, which suggested that with normal aging, elevated blood pressure was necessary—hence the term "essential hypertension"—to supply enough blood to organs, especially the kidneys. White addressed this misconception as early as 1931 in his textbook on heart disease:

> The treatment of hypertension itself is a difficult and almost hopeless task in the present state of our knowledge, and in fact for aught we know, the hypertension may be an important compensatory mechanism which should not be tampered with, even were it certain that we could control it. Years ago there existed a simple rule for the height of the systolic blood pressure (the upper number) and frequently even today one still finds this erroneous rule accepted. This rule stated that the blood pressure should be 100 millimeters of mercury plus as many millimeters as the years of the individual's age.[16]

So a blood pressure of 160/80 would be acceptable for a sixty-year-old, and 170/85 for a seventy-year-old. Instead, for all ages, White recommended a range for systolic pressure of 100 to 145, and a usual range of diastolic pressure of 70 to 85, numbers that are remarkably close to those we consider acceptable today.

Yet physician practice is notoriously slow to change, and for decades longer, doctors would continue to consider the inching up of blood pressure as a harmless sign of aging.

Animal surgery began to challenge this dogma. A French physiologist, Claude Bernard, slit the sympathetic nerves to the ear of a rabbit and discovered that the blood vessels in the ear dilated.[17] By 1913, French surgeon René Leriche pursued that clue and tried cutting sympathetic nerves to limbs in patients with poor blood flow to their arms or legs.[18] By 1925, a team of surgeons at the Mayo Clinic operated on a patient with high blood pressure, severing a segment of the sympathetic nervous system. Following sympathectomy, the patient's blood pressure fell, and his heart and kidneys continued to function as before the surgery.[19] The experiment was one of the first pieces of evidence that lowering blood pressure at least did no harm to vital organs.

For a time in the middle of the twentieth century, sympathectomy remained a viable procedure for lowering blood pressure. The nerve-severing technique was reserved for those with very high, or malignant, blood pressure. But it was a clue, giving researchers reason to believe that high blood pressure was not inevitable or "essential," and that it could be treated without doing harm—perhaps some day with drugs rather than surgery.

There were attempts before the Framingham Heart Study to find factors that promote risk for heart disease. As early as 1937, Massachusetts cardiologists, including Paul Dudley White, Howard B. Sprague, Samuel A. Levine, Edward F. Bland, and Menard M. Gertler, compared one hundred young coronary heart disease patients—all under the age of forty—with healthy people, looking for clues by considering how the two groups differed. In the search for clues, two things stood out. The first was that there were ninety-six men and only four women in the coronary group. The second was that the majority of patients were husky, short, and thickset. The nation's preoccupation with the war precluded further study of this group of patients, but the cardiologists had several years to ponder the relations of male gender and mesomorphic build to heart disease.

As soon as postwar funds became available, the Boston specialists began work on the Coronary Research Project. In 1946, White and his team recruited another one hundred young patients for their project, this time ninety-seven of them men. White was convinced that heart disease in young men was, in large part, their own fault and not a random act of fate. And he believed that with heart disease, unlike polio or measles, no single microbe lurked, no lone virus lay in wait. Instead, coronary disease developed from a multidimensional attack of several unrelated factors. The goal of the project was to identify those causes.

Menard Gertler, then executive director of the Coronary Research Project, pursued the notion that medical science could predict who might be at risk for heart disease by investigating two prime suspects—high blood pressure and high blood cholesterol. At Massachusetts General Hospital, researchers began to profile the coronary-prone man—still assuming incorrectly that heart disease was exclusively a male problem.[20] At the start of their research, the Public Health Service was reporting that fully half the deaths in the United States occurred because of cardiovascular-renal diseases—more than the next four causes combined: cancer, accidents, tuberculosis, and pneumonia. The volunteers had already suffered a heart attack. They were ineligible for the study if they had diabetes or hypertension, an unfortunate design flaw that would prevent the scientists from identifying these conditions as promoters of heart trouble. The coronary patients were compared with healthy volunteers from a nearby plant, Lever Brothers in Cambridge. The researchers concluded, in part, that there was "little doubt that 'pure' coronary heart disease (that is, uncomplicated by hypertension or diabetes) is predominantly a male disease prior to the age of 40." They were correct for that age range, but may have inadvertently set back research on heart disease in women by the common interpretation that cardiovascular disease was solely a male concern.

These early attempts showed a growing understanding of the enemy's complexity. The scientists uncovered some likely culprits, such as obesity and family history. They identified 32 of the healthy controls as "probables," meaning patients who were likely to develop heart disease. These were people who felt fine and had no reason to suspect that health problems were in their future. Yet within four years, 28 of the 32 probables developed heart disease, and 20 suffered heart attacks, including five fatal ones. But the researchers suspected some factors, as well, that turned out to be innocent, like thick wrists, a mesomorphic build, and the season of winter. Alcohol use was found to be neither risky nor protective. At least one major culprit, tobacco, which would be proved guilty beyond doubt twenty years later, escaped detection this time around. The scientists were studying those who were already sick with heart disease and then trying to trace the origins backward, a process easily tripped up by faulty memories and an inability to separate cause from effect. But they learned enough for White to proclaim: "Heart disease before forty is our own fault, not God's or Nature's will."[21]

White had been thinking about heart disease as a preventable illness ever since he brought the electrocardiograph to Boston in 1913. He tried in the Coronary Research Project to tease out an understanding of causes by looking at patients who already had suffered heart attacks. While his study's success was limited, he knew what the next step must be. He became convinced of the value of a well-designed examination of healthy people to see who developed cardiovascular disease and who did not. It was a vision of what would become the Framingham Heart Study, and that vision would motivate White to advocate and later use his enormous influence to lobby for a prospective epidemiologic study. For the rest of his life, he remained a staunch and influential supporter of the Heart Study.

These were the clues that postwar medical researchers had at their disposal in their search for the underlying causes of heart disease. They believed that cardiovascular disease had something to do with diet and with cholesterol in the blood. Few appreciated the role of hypertension. And while some researchers were beginning to suspect tobacco, others still smoked their way through study design discussions. Just as technology had revolutionized the diagnosis of heart disease with the electrocardiogram, thousands of innovations would follow, most importantly computers. Today, we are accustomed to technology racing ahead of law, of ethics—even of the human mind. But at the midpoint of the twentieth century, technology was usually a step behind the curiosity of scientists.

By 1948, scientists who would design the Framingham Heart Study knew little about what they'd find, but a lot about the questions they needed to ask. They had threads of research going back 150 years, put together by pathologists, clinicians, and laboratory scientists. These disparate threads were barely beginning to be woven together into something that might resemble a coherent set of hypotheses about the causes of heart disease. For the next half century, the scientific dialogue would continue to feed their research, and in turn be nourished by them. Scientists and physicians had technology such as the electrocardiograph to study the disease, but they couldn't have known the full extent of the technological revolution that would aid and accelerate their work. A side effect of the space race that put a man on the moon in 1969 would give them high-speed computers to enable them to understand the gold mine of information they were accumulating.

Although they were not aware of the extent to which their research would change, the Framingham Heart Study's designers seemed to count on the steady advances in science and technology almost as articles of faith. Fifty years later, two Framingham pioneers, Roy Dawber and William Kannel, recalled the naïveté mixed with optimism that got them started. "We were about to study 5,209 people and eighty variables. Some thought we were insane to try to do this. And I think it turned out we were," says Kannel. And in wry understatement, Dawber adds, "After a few years, it was obvious we needed a little more training in epidemiology."[22] They knew that the volumes of information they were about to gather would fill storage areas with overflowing boxes and file cabinets. They understood the limitations of their paper-and-pencil records. They realized that their slow, clumsy calculating machines would never keep up with the exponential growth of information. Like most physicians, they knew little of statistics, apart from a few basics. They were aware that they were about to plunge into something that over time could overwhelm them.

Armed with the observations of Heberden, Herrick's hypothesis, and the machinery of Einthoven; aware of the clogged arteries of Anitschkow's rabbits and of Bloch's explanation of the action of cholesterol molecules; intrigued by the theories of Keys; inspired by the work and dream of White; and backed by the U.S. Public Health Service, researchers were ready to launch their own visionary and ingenious scheme in Framingham.

They were physicians who knew full well how much they had yet to learn. But they could have had little idea that their efforts would ultimately alter the focus of the entire medical profession, tilting it toward prevention. The Study's findings, now so patently obvious to almost everyone, represent a miraculous shift in thinking over the course of half a century.

FOUR

A Struggle for Identity

Gilcin Meadors was a young, enthusiastic physician with a Southern drawl and a boyish face. He was fast to retort with a self-mocking sense of humor, quickly conceding to his new Northern neighbors that while his was the side that lost the Civil War, his was also the side that knew how to pronounce its *r*'s. His speech was as strange to them as theirs was to him. Nell McKeever was a public health nurse from Oklahoma, speaking in the slow, square syllables of the Southwest and patiently teaching the citizens of Framingham that "epidemiology" meant the study of the occurrence and distribution of diseases in populations, and then explaining to them how researchers hoped to use the science in a new way. They were a personable team, the first representatives from Washington, D.C., to begin work in the town of Framingham for the Heart Study.

The federal government had made a commitment to fund a study looking for the root causes of heart disease. It was clear by 1948 that the right study had to be done, but exactly how to fashion it was still far from clear. Framingham, Massachusetts, was, in the opinion of leaders at the U.S. Public Health Service, the perfect place for it, a town representative of latter-day demographics.

There were other villages, towns, and cities to consider because they fit a similarly typical mold. The chief rival site for the study was Wilkes-Barre, Pennsylvania, a town picked by Bert Boone, who was a surgeon at Temple University and one of the Study's designers.[1] Then, at a medical meeting in Atlantic City, New Jersey—where doctors of that era claim to have learned more on the boardwalk than in the meeting halls—he met Dr. David Rutstein, a Harvard classmate who grew up in

Wilkes-Barre. "Rutstein said, 'No. No. No. Not Wilkes-Barre. The doctors there won't go along with it. They'll think it's socialized medicine. And you'll be nowhere near a medical school. I'm going to help you find a town in Massachusetts,' " remembers William Castelli, who studied under Rutstein.[2] Rutstein was heading off to Harvard to become a professor of public health. He convinced Study organizers in Washington that Massachusetts was the right place for their "experiment" in public health. Other Bay State towns were considered, including Beverly and Peabody. But Beverly was already slotted for a state program to address childhood rheumatic fever, and a second study there would overtax residents' resources. Peabody was eliminated because it was "too far from Harvard," which even then figured prominently in medical circles. Framingham, on the other hand, "is closer to the residences of the Boston cardiologists, which [Rutstein] seemed to think would make a difference in their attendance," according to early planning documents.[3]

In the late seventeenth century, Framingham was a haven for those fleeing witchcraft hysteria in Salem. It was a farming community, home of the first teachers' college and of the first women's prison. Later, it would house one of the nation's first shopping malls, with fifty stores and free parking for six thousand cars.[4] Framingham was a market researcher's dream come true. If the broad middle-class population of this town liked something, it would sell throughout the country.

Rutstein liked Framingham, and his legendary take-no-prisoners style is often credited as one reason the Heart Study found a home there. Rutstein wanted it to be close to Harvard, which was becoming the cardiology center of the country, and he was not shy about pushing for it. "He was a little bit on the aggressive side with everybody," says William Zukel, then associate director for collaborative studies at the National Heart Institute.

Behind the scenes, but no doubt equally persuasive, was White, a man with a softer side. By 1948, more than three decades after he brought the electrocardiograph across the Atlantic, White had emerged as the leading cardiologist of his day. He had recently been named executive director of the National Advisory Heart Council and was chief medical adviser to the National Heart Institute.

He was an impressive and influential figure, but not an intimidating one. "He never lorded over any physician who invited him to consult on a patient. He was very positive in his assessment of what the physician was already doing, but would then give his advice as to what he thought should be done for optimal treatment," says Zukel. White also traveled widely, speaking in hospital basements or anywhere there was space, about emerging knowledge on heart disease. As his reputation grew, White's speeches would recruit physicians from all over the state, from across the country, and eventually from around the world. Once, when White was on a speaking engagement in North Dakota, physicians from a six-county area were invited to Grand Forks, where he was talking about lifestyle and heart disease. It was a cold February Sunday morning—the kind of day made for staying in bed. Every physician registered in the state medical society from the six counties had been invited, though Zukel, who helped organize the event, was skeptical about the number who would actually make the trip. He believed physicians were uninterested in heart disease. But 100 of the 135 invited physicians came—and the room was even more packed than that. "It turned out that more doctors showed up than were registered in the area. We found out later that doctors in Canada heard about it, and came over the border to hear him speak," remembers Zukel.[5] The assumption that doctors weren't interested had been false. It was more likely that they were frustrated by being unable to intervene with their heart patients in any meaningful way.

White set a tone of collaboration, not competition. It was just the tone that the health sector in Framingham needed to feel reassured that the Heart Study was not going to take its patients or attempt to control its practices.

White was a trusted adviser and close personal friend of the National Heart Institute's first director, Cassius Van Slyke. His reputation as a cutting-edge researcher, an excellent diagnostician, as well as a modest and kind human being, cannot be underestimated in measuring his influence in bringing the Study to Massachusetts.

Some thought that Framingham, with a population of 28,000, might be too small, requiring the cooperation of nearly half the adults in the Study age group. But apart from that, it was just what was needed. "It was not just a suburb, it was an independent town. He [Rutstein] got some reassurance from the medical society in Framingham that they would be receptive to doing this. That was important.

It was a time in history when there was some talk of socialized medicine. Doctors didn't want the government involved in their practice," recalls Dawber, director of the Study from 1950 to 1965.[6] Pioneering researchers had to tread lightly.

At the time that national scientific experts were examining the town's statistics, the median family income in Framingham was $5,277. Only ninety-three families were truly wealthy, earning more than $25,000 per year. Framingham residents lived in 7,105 homes, and 4,066 owned them.[7]

Most citizens worked in the self-contained town, including teachers at Framingham State College and doctors who had privileges at Framingham Union Hospital. Lawyers had enough work to sustain practices in town. A few commuters would board the train to Boston. As in other New England towns of the mid-twentieth century, the borders of Framingham went beyond the residential area and into the surrounding farmlands. Marion Kittredge, now in her nineties, was not unusual in growing up with chores that included tending the chickens, pigs, and cows. But mostly, it was a factory town whose workers produced baggage labels, paper products, glue, and rugs. "Most people walked to work, to the Dennison Manufacturing plant, or Hodgeman Rubber or the Elizabeth Carpet Factory. The Dennison whistle would blow about five minutes to eight and you knew it was time to get to work," says Rita Duran, who lives in neighboring Natick, five miles from where she was born.[8]

The townspeople ate an ordinary diet, earned an average income, worked in a typical variety of jobs, smoked as much as most Americans, lived in run-of-the-mill houses, and fell victim to the usual variety of diseases. Their ethnic mix—including roots in Poland, Ireland, Italy, Greece, French Canada, and England—was a reflection of European ancestry in America at the time. Nearly half a century later, the Study would recruit minorities to reflect the changing diversity of Framingham.

They were perfect for the Study because of their very contentment, a civic satisfaction that would lead researchers to believe that they had every reason to stay put and live out their lives within the twenty-six-mile-wide town or on the dairy and vegetable farms that ringed it.

The Study's researchers have never lost track of exactly who was doing whom the favor. From the beginning, it was crucial to get the townspeople's support so that they not only signed up, but came back every other year for at least twenty years, or until they died.

Framingham residents had already proved that they could cooperate with medical science. From 1916 to 1923, five thousand residents volunteered for the Framingham Tuberculosis Demonstration Study.[9] Walter Sullivan was a small child then, born in 1914. "I remember that TB was a dreaded thing. People were snatched out of the community and segregated. I can remember going with my father to a hospital outside of Worcester and we'd wave to his friends through the windows." Framingham was chosen for the TB study during World War I because it was average even then in size, in the amount of overall disease, and in the number of cases of tuberculosis. Researchers found unsanitary conditions in a neighborhood called Saxonville and poor hygiene and overcrowding in nearby Coburnville. In a preventive effort, about 6,600 residents were examined for tuberculosis. By the end of the study, physicians were finding the disease sooner—75 percent of cases were reported in the early stages in 1925, two years after the study's completion, compared with only 45 percent found in 1916. Early treatment and improved sanitation led to a 55 percent decrease in cases of tuberculosis and a 6 percent drop in the death rate. There were not yet effective antibiotics for the disease, and treatment was simply good nutrition, rest, and fresh air. Sewer systems in town were improved, and farmers were taught hygienic farming and dairy practices. The study was an important milestone.

Now, the sons and daughters and neighbors of the TB study volunteers were going to be asked to reveal their daily habits, the chemical makeup of their blood, and, ultimately, the causes of their deaths, in an effort to pinpoint the culprits behind the epidemic of heart disease.

But as with many monumental moments, the start-up passed largely without fanfare. The locals had more immediate concerns. A coveted factory was coming to town that would provide stable jobs and a surge in population. During the war, the big automobile plants in Detroit stopped production of passenger cars and shifted to wartime needs. But in 1945, General Motors bought a 126-acre tract of land on the edge of Framingham, and in 1948 it began producing Buicks, Oldsmobiles, and Pontiacs. The plant eventually employed 3,500 workers who covered two shifts.[10] Many of them came to New England from the Midwest.

Shoppers World arrived in town in 1948, too, breaking ground on Route 9 and making the nearby tree farm owned by Richard Wyman, where generations of Framingham residents picked out their Christmas trees, suddenly worth a fortune. Shoppers World was one of the country's first malls. Route 9, like highways all over America's suburbs, became flanked by brick and mortar and paved with asphalt.

Entertainment was becoming passive. The *Framingham News* reported on June 10, 1948, that television made its debut in town. For $349.50, families could buy a ten-inch black-and-white Philco at the Lewis Furniture Company on Irving Street. Soon, roller rinks weren't as busy as they once were, and children had to be coaxed to leave the living room and go outside and play. Fewer adults went dancing, although dancing had been a staple of entertainment during the Depression and even the war years. Across the country, grand old dance palaces like the Aragon Ballroom in Chicago and the Meadows in Framingham went dark as more and more cathode-ray tubes lit up living rooms.

Eventually, even the downtown movie theaters—in Framingham, the Loew's Hollis, the St. George, and the Gorman Theaters—couldn't keep up with the convenience of the growing number of cinemas at strip malls with easy parking. All across America, supermarkets were popping up. The Jewel Foods, A&Ps, and Safeways of the country were among the first fully air-conditioned public spaces, and their openings were townwide celebrations.

Downtowns were beginning to die. "Downtown Framingham used to be a mecca for shopping. People from Ashland, Hopkinton, Natick, and other towns would come to Framingham to shop downtown. We had all kinds of offerings. There was a Gilchrest, a fairly large department store, and Kresge's and Woolworth's and Newbury's. We had small markets. Everybody came to Framingham to shop," says Sullivan. "We had three movie houses downtown. You'd go to a restaurant, see a movie, and then stumble home. All of that changed when Shoppers World came."

The year 1948 was memorable for baseball fans in Massachusetts, one of those heart-stopping, near-miss years that lives on in the memories of Boston fans, perennially disappointed until their 2004 World Series triumph. The Braves dominated the sports pages, and they were contenders in the World Series. Framingham native Fred Hoey did the play-by-play over the radio. Two pitchers—left-hander Warren Spahn and right-hander Johnny Sain—were the team's backbone, and fans crossed their fingers and hoped through the late summer that a pennant would come to Boston.

Everything was growing: jobs, housing, shopping centers, and families. The beginning of the Heart Study was, by comparison, a nonevent. When Mary Sullivan, a Study nurse, first stepped into the examining room on September 28, 1948, as a test-run volunteer, the *Boston Globe* ran a story proclaiming heart disease "the worst enemy of man on this earth, and medical science is mobilizing to do something about it." But the report had nothing to do with the opening of the Heart Study less than twenty miles away. The article concerned a plan by the American Heart Association to expand its network of associations around the country. Both locally and regionally, the news of this effort to understand heart disease was overshadowed by an announcement of the formal opening of the GM plant in Framingham with its promise of job security for thousands.

On October 11, the day the Heart Study officially opened its doors to volunteers, the *Globe* printed an aerial photograph of the Framingham auto plant showing a million square feet of space that would employ two thousand workers—72 percent of them veterans. It would bring in $66 million a year to New England. That day, with five hundred business leaders in attendance and buses hired to shuttle the town's residents to the plant for tours, GM president Charles Wilson cut the ribbon to the plant that would soon be rolling out eighteen automobiles an hour.

That day was also Eleanor Roosevelt's sixty-fourth birthday, and her syndicated column, filed from Paris, concerned the leisurely pace of dining in France. "The French cannot understand why we hurry over meals," she wrote.

But the headline that Framingham residents most likely remember from that autumn is "Cleveland Wins World Series." More than 40,000 fans filled Braves Field, most waiting hours—some even days—in line for tickets. The Gillette Safety Razor Company set up numerous television sets in Boston Common so that thousands more could cluster around and watch outside as the Braves lost the World Series in a deciding seventh game.

The announcement of the opening of the Heart Study was a small item buried in the local newspaper.

GILCIN MEADORS and Nell McKeever, the Heart Study's first representatives in Framingham, won the public's trust with diligent education and by allaying fears about privacy and government intrusion. "Nell enlivened us so that we did what she wanted us to do. Through her, I first heard this "epidemiology" word. She'd say it and string it out a little: ep-i-de-mi-ol-o-gy," recalls Walter Sullivan.

The researchers also became part of the community. "They were hard workers, but Meadors was a dandy relaxer and played the banjo. We had more fun with Meadors and his Mississippi music. He had a wonderful personality but a serious approach to his job. He and Nell were warmhearted. These two didn't step on anybody's toes; they didn't offend anybody. Their manners were such that you didn't mind listening to them and trying to do what they asked you to do. Off hours, we'd gather in our house, Katie [Sullivan's late wife] and groups of friends," says Sullivan. One of the guests and Study participants was the school's music director. Between the music director's singing and Meadors's banjo, it was a good way to build camaraderie and trust among the town's leaders.

Those that Meadors and McKeever encountered had the same opinions and fears that permeated American society in the late 1940s and early 1950s. They were unaware of the risk factors for heart disease. It was a big job, therefore, to convince them that enrolling in the Study would aid medical advancement.

Meadors and McKeever were swept up by the momentum of a scientific community and eager to study the epidemic. But the people of Framingham had other medical problems on their minds. They were still worried about polio, though in 1947 the number of cases in the United States had dipped to 8,000, down from 25,000 just the year before. The drop was due not to the appearance of the Salk vaccine— that didn't become available until a few years later—but rather to the vigilance of worried parents who kept their children away from public pools and beaches.

Health columns of the day talked about toning exercises and calisthenics geared for "older" people in their thirties. Food pages gave hints on easy meal preparation. None of it had anything to do with heart health. In fact, there was scarcely a mention of heart attack or stroke outside the obituary pages.

Meadors's first order of business was to attract attention. He went to town meetings and recruited volunteers from PTA sessions, churches, and civic organizations. He looked for the leaders in town, and he befriended them. He talked about health and the heart and life and death. But he also socialized with civic leaders, who came to understand the importance of the mission. They were ready to spread the word.

The citizens responded to the Study's openness and inclusiveness. Victor Galvani was a member of the town board then, and his reaction typified the responses to the coming Study. "I was excited because no one saw doctors generally, let alone specialists on heart studies. It was exciting that they were going to do these sophisticated tests, and all for no charge."

That was one aspect of Meadors's message: Join the Study, and there is something in it for you. But there was a more important theme that began in the first days and grew stronger over the years of the Study: Sign up and be part of an effort that will make an enduring contribution to humankind.

Transforming a community that might resist medical intrusion to one that would take pride in the Study was the job of local volunteers who knew their town and its neighbors. Meadors's major contribution during his short time with the Study was identifying leaders and sending them out as emissaries. Their mission was to help researchers round up five thousand subjects willing to participate in the Heart Study. Some volunteers knocked on doors; others staffed the telephones and dialed every other number in the local listings. Back in the days of rotary dials and exchanges like FRamingham 2, citizens became telemarketers, talking up the Study. "You needed someone who was good at phone sales," says Dawber of the effort that continued into 1949, when he took over the Study. Still others were making presentations before school committees, at town meetings, before groups like the Veterans of Foreign Wars and the Christopher Club Society. They had scripted responses to anticipated questions.

"Yes, the results will remain absolutely confidential."

"No, Study physicians won't treat participants. They will send all results to a family physician."

"Yes, it will be completely free of charge."

The goal was to persuade five thousand healthy adults to volunteer, even knowing they would not personally benefit by joining. In fact, early predictions were that 400 of them would suffer cardiovascular disease by the end of the fifth year of the study, 900 of them by the end of the tenth year, and 1,500 by the study's scheduled end, twenty years later. At the time, researchers were looking to induct fully half the town's adult population between the ages of thirty and sixty, an unprecedented recruitment effort in both raw numbers and percentage of a population studied.

By the Study's first anniversary, the *Framingham News* reported that more than 2,000 had been examined, and appointments were scheduled for another 1,000. "While the director of the program would like to take everyone who volunteers, something had to be done to limit the number of examinations. This has been accomplished by limiting the age of participants to the 30 to 59 year group, and by the use of a representative cross-section of the population. The lucky ones whose names are 'picked out of a hat' will be invited to have the heart examination and thus make their contribution to medical research."

Meadors got in at the starting point of a Study that everyone agreed was needed. But there was little agreement about exactly what to include in the project. The Study didn't have a clear mission, and he didn't have a clear charge. No one before had done what he was setting out to do, designing a study using a townful of ordinary people to unravel the mysteries of multifaceted, highly complicated diseases. Even the best minds in the country who advised the Study began with notions that were ill-conceived. They would need a year or two for clarity to set in.

Meadors was an administrator and a bureaucrat. He had skills that helped propel the community into action. His initial instincts in designing and steering the course of the effort were on target. In an outline of his plans dated July 19, 1947, he wrote:

This project is designed to study the expression of coronary artery disease in a "normal" or unselected population and to determine the factors predisposing to the development of the disease through clinical and laboratory examinations and long term follow-up of such a group. A complete history will be taken on each individual, and his cardiovascular status will be appraised by means of a thorough physical exam and adjunct lab procedures including X-ray measurement of the heart, electrocardiographic and electrokymographic tracings and chemical study of the blood.

While his research qualifications were all but nonexistent, he was eager to begin and anxious to get settled in Massachusetts before winter set in. He moved from Washington in September 1947, and accepted an offer from Rutstein of workspace at the Harvard Medical School.

Meadors and McKeever went about the work of getting an office up and running. They ordered pens and inkwells, desk blotters, typewriter ribbons, telephone index cards, and a three-hole paper punch. The stationery was imprinted "A record of your heart," with a logo of a black phonograph record whose center hole was in the shape of a heart. They bought a Plymouth for $1,300 and budgeted $25 a month to operate it. For the smokers on staff, they ordered three ashtrays. They also obtained equipment that represented the best that technology offered at the time: a fluoroscope and a Simplitrol Electrocardiograph by Westinghouse. They purchased a Friden automatic calculator, a machine that, lacking the capacity to do automatic multiplication, could not possibly keep up with the work that lay ahead.

They also acquired an electrokymograph from Cambridge Equipment Company,[11] a new piece of technology that would quickly prove useless and today sends middle-aged and younger cardiologists to the internet to figure out, in vain, what in heaven's name it might have been. The theory behind it was that as the heart beats, it enlarges with each inward rush of blood and diminishes in size as blood is expelled. There is a contraction of the edge of the heart shadow that shows on an X-ray with each beat. At any particular point, the shadow would move in and out. The electrokymograph translated that motion into a linear image on a screen. Hindsight is clear on this innovation. "As a way of detecting heart disease, it was about as ridiculous an idea as anybody could think of. It never turned out to be of any value. But at that time, there wasn't much else. They were grasping at straws," according to Dawber. At the fiftieth anniversary celebration of the Study, Dawber explained that the original attempt to screen for disease was "somewhat ridiculous."

The Framingham Study began as an effort by the U.S. Public Health Service, with assistance from the Massachusetts Department of Public Health and advice from the Department of Preventive Medicine at Harvard Medical School—institutions concerned primarily with disease prevention and control. David Rutstein used his considerable influence to reconceive the Study, placing greatest emphasis on early diagnosis and relegating epidemiology to the back burner.[12] He became a highly influential adviser to the project, enjoying the collegiality of the leading cardiologists of the time, including Paul Dudley White and Samuel Levine, who also served as advisers.

Rutstein was not satisfied with the qualifications of Meadors, who proved to be out of his element. Meadors came to New England with cultural baggage—a Southerner among Northern academic elites, a government bureaucrat among independent physicians, and a medical administrator with limited clinical experience among cardiologists. Yet, like any physician, Meadors wanted to help figure out how to prevent and control the mysterious epidemic of heart disease and stroke.

Heart attacks were occurring in record numbers, and few survivors returned to any semblance of an active life. A common bond among everyone involved in developing the study was determination. Long pent-up ideas were finally given opportunity for expression. Scientists could at last attempt to solve the mysteries of heart disease. In 1948, in the face of medical ignorance and powerlessness, the first uneasy step was to agree on the correct questions.

To identify the causes of the disease, it was crucial that there be long-term observation of healthy people who had never before suffered a heart attack or stroke. The approach was critical in order to note what went wrong, and in whom. To maintain the trust of local physicians, the researchers in the clinic would not treat or even offer advice to the volunteers they were seeing. They would use these citizens, and their heart attacks, to begin to trace the clues backward, eventually to levels of blood pressure and cholesterol, to diabetes, obesity, amount of exercise, and tobacco use. But following that trail would take years. The project would require experience, and no one had experience in anything resembling the proposed two-decades-long epidemiology project. And it would take persistence. The project would mean that those involved would have to gain experience quickly and learn patience and perseverance.

But the human urgency of seeing thousands die before their time tested the Study designers' patience and tempted them to try to prevent heart disease. The study became a tug-of-war between two lofty objectives. One was to be a long-term observational study of healthy people, which, indeed, prevailed. A second was to diagnose and prevent heart disease in the community—the traditional public health approach backed by Rutstein. An early written aim was "To reduce disability from heart disease and to delay the onset of cardiovascular disease in the older age groups of the population."[13] While that looked great on paper, the truth was that no one knew how to accomplish those goals. To try to prevent heart disease at a time of profound medical ignorance was a fool's errand.

The second objective—to *control* cardiovascular disease—eventually fell by the wayside. Ultimately, it was not a task the doctors and researchers of Framingham, or of any other community, could address until scientists first unraveled the causes of heart disease and stroke. It evolved into a small program called the Cardiovascular Hygiene Demonstration Project, based in nearby Newton, Massachusetts, and led by Dr. Lewis Robbins. The purpose of the Newton program was to determine how existing knowledge of prevention, diagnosis, treatment, and rehabilitation of cardiovascular disease could be applied within community health programs. It couldn't offer much, but arguably some applications from the program could be tried. In the spring of 1951, a booklet on what was then called the Newton Heart Demonstration Project outlined modest efforts: "Monthly lectures on heart disease; nutrition activities including the booklet *Planning Low Sodium Meals;* a pilot study on group technique in weight reduction; a trial class in weight guidance for overweight high school students; prevention of rheumatic fever estimated to eliminate one-third of future valvular heart disease by preventing strep infections; weekly throat cultures taken in schools on a group of 38 rheumatic fever children who are receiving daily oral penicillin."[14]

The Newton program's biggest success was probably in helping to prevent recurrences of rheumatic fever in children. Youngsters could recover from the fever, but a second or third bout, known to be caused by streptococcal infection, could permanently damage their heart valves. "A second attack could be prevented with penicillin, but it was very expensive for families to have a youngster on daily penicillin," says William Zukel, who started his career with the U.S. Public Health Service as a physician with the Newton study. So Dr. Ernest Morris, a state health official in Newton, contacted the Pfizer drug company and got it to agree to provide the medication in bulk, as part of a prevention study, at a substantial discount—about ten cents a day, compared with the market price of twenty-five cents a day. "It was one of the earliest attempts by the Public Health Service to learn about prevention of heart disease," Zukel says.[15] The Newton study also worked with local dentists, teaching them to give penicillin to patients with valvular or congenital heart disease before they underwent dental procedures as a way to prevent a heart infection called bacterial endocarditis.

But the Newton program could not begin to scratch the surface of the far more prevalent epidemic of heart attacks and strokes. It would take a couple of years to sort out the relative importance of the two objectives, to come to the harsh realization that before physicians could *do* anything about heart disease, they had to watch and bide their time. The *doing* would have to wait for others who would take the clues soon to be identified in Framingham and use them to fashion treatments and preventive measures. Even now, more than fifty years later, those preventive measures are falling frustratingly short of helping all those in need. Had the Framingham Heart Study continued on a path of pursuing both objectives—observation and control of heart disease—it might have disappeared into oblivion, the fate of the Newton study.

Meadors and McKeever were in on the ground floor of a study whose goals were conflicted and unattainable. The long-term success of the observational study depended on community mobilization. And that's where Meadors made his mark, laying the groundwork among the citizens, winning their support, convincing them that, by agreeing to a thorough medical examination every other year for twenty years— or more—they would leave a lasting legacy.

News of the Framingham Heart Study was spreading enthusiastically, from town council member to citizen, from employer to employee, from neighbor to neighbor. Newspaper articles, radio announcements, and civic meetings got the news out, but it was word of mouth that persuaded people to put their fears to rest. The participants, along with their personal physicians who believed in the Study's goals, had gotten the reassurance they needed most: an ironclad promise of confidentiality. "When you told them they were going to keep a record and make a study of it, people first said, no, they did not want to join. I kept assuring them that their records would be private. I really had to sell them. But it was my neighborhood. I knew these people. I sat down and talked with them," says Evelyn Langley. With efforts like hers going on all over town, her neighbors became interested in this high-level medical appointment.

The team began scheduling appointments. On September 29, the day after Mary Sullivan made her test run through the maze of examination stations, the local doctors in town put on hospital gowns, blew into a tube to have their lung capacity measured, had their blood pressure taken and their blood drawn, and exchanged roles with patients, if only for a few hours. Meadors wanted them to go back to their patients and be able to tell them, firsthand, what the examination would be like. Within weeks of the Study's official opening on October 11, the staff had filled appointments through February 1949.

But Meadors's lack of administrative skill was straining both him and the Study. File folders within the U.S. Public Health Service were filling up with letters of both complaint and support from scientists, from staff members at the Heart Study, and from physicians about the competence of Meadors to run the project. He had backers from the local medical community, opponents at Harvard Medical School. He had the strong advocacy of citizens who became friends, like board member Walter Sullivan, and staff people like Nell McKeever, but the condemnation of the staff statistician, who quit in protest of his lack of administrative skills. By the end of 1948, he and some of his staff were referring to office operations as "this mess."[16]

It was an era when doctors all over the country were feeling uneasy, and local Framingham physicians were no exception. With talk of national health insurance, supported by President Truman, doctors feared the advent of socialized medicine. The bureaucrat Gilcin Meadors was hardly the man to reassure them.

The postwar era was the heyday of physician independence, of autonomy and respect for the medical profession. Medical knowledge was on the verge of exploding. In this climate of freedom and independence, doctors were wary of a system of government control. They needed a lot of assurance that the government wouldn't be looking over their shoulders, and they worried that the Heart Study might be the first step toward broader federal intrusion. They also wanted to be sure that they wouldn't lose their patients to the Study's doctors. They had to have absolute trust in the leader of the Heart Study. Meadors's background as an administrator made them jittery. His accent and his ties to Washington reminded them repeatedly that he was not one of them.

Shortly after Meadors began work, Washington's control of the Study changed hands, going from the Public Health Service to the National Institutes of Health and the newly established National Heart Institute. The timing was right. The new institute had the scientific capabilities to create a study design and analyze data. A memo to Meadors dated May 13, 1949, from Dr. A. L. Chapman, chief of the Division of Chronic Disease of the Public Health Service, gave notice that the very transfer of the Study to the NIH indicated a major shift in focus, away from an attempt to make it a preventive program. "[T]he blunt fact could not be avoided that you are doing an epidemiology study the prime purpose of which could be more logically classified as research rather than service." On August 30, National Heart Institute director Cassius Van Slyke wrote to Meadors, again clearly outlining the change in the Study's mission. It was now redesigned "to correlate the findings with respect to the development of degenerative cardiovascular disease in a cross-section population with the finding of clinical examinations prior to onset of the disease, in order to detect early signs pointing to probable development of disease, and where possible, to discover etiological factors." In other words, watch a healthy population for a long time, wait for them to develop heart disease and stroke, and then look backward for the clues—later dubbed risk factors—to what made them sick.

The move to the National Heart Institute from the Public Health Service and the shift in mission from prevention to observation clarified the core objectives of the Heart Study. It also sealed Meadors's fate. Now, Heart Institute scientists were heavily involved, and Public Health

Service bureaucrats were not. The discipline of science, with its hypotheses, methods, and patience, was taking center stage. No longer would the Study be under the control of the service charged with preventing disease and improving public health, an unattainable mission when so little was known about cardiovascular disease. It had moved to the nation's scientific research center, and Meadors lost his footing. The Framingham Heart Study was destined to become a jewel in the crown of the world's foremost medical research institution, and the calculation was that it would take twenty years.

And so Meadors moved on to a new assignment with the National Cancer Institute. Eventually, he would work in private practice in Maryland. He left behind a generous reserve of goodwill among citizens of the town. His greatest legacy was the foundation of support from the community that he and McKeever laid with the strength of their personalities.

Robert Sullivan, part of the second generation of participants in the Heart Study, was a boy of about six when his father, Walter, became active in the citizens' committee for the study. He has fond memories of Meadors and McKeever coming over to the house. His father has nothing but praise for the early efforts. "They did all this legwork and groundwork. Nell McKeever worked her head off setting the thing up and getting this whole group of people involved. They did stuff that was both medical and nonmedical," says Walter Sullivan. "Meadors was a true Southerner and a real great guy. They were two characters that I learned to love as people. We became fast friends."

Meadors left behind a community willing to be studied. He and McKeever had persuaded the citizens of Framingham to reveal their habits, histories, and body measurements, and to donate their blood, with no promise that it would directly benefit any of them. Perhaps his greatest achievement during his brief tenure was setting up a network of key community leaders, known and trusted by their neighbors. They were veterans, lawyers, housewives, and parents who would spread the word that the Heart Study was an endeavor worthy of the town's participation. Thanks to Meadors's efforts, the next director of the Heart Study, Roy Dawber, could start his work with a groundswell of goodwill and cooperation.

FIVE

The People Who
Changed America's Heart:
Voices from Framingham

The worst of it for Evelyn Langley was getting a door slammed in her face, as if she were a common saleswoman, or worse, a scam artist. She was part of the foot patrol for the launching of the Framingham Heart Study in 1948, and her assigned area was the ten blocks surrounding her local elementary school. No one knew then that those with the good fortune to live long enough would be in for more than half a century of meticulously crafted medical examinations. It was rare for a neighbor to shut the door on her, but when it happened, it always hurt her feelings. She walked up and down the blocks, knocking on one door, skipping the next, knocking on the third door, until she hit every other house in her area. These New Englanders had a well-guarded sense of privacy. They weren't used to going to doctors. They were worried about a thorough examination that might find an ailment they'd prefer not to know about, and probably could do nothing about even if they did know. They didn't want to be poked and prodded by doctors doing research. Mostly, they didn't want anyone, especially the government, to get too familiar with the medical details of their lives. "Sometimes you practically had to beg people to go on the Study," says Langley. "Now, they'd break your arm to get in."

In 1948, Langley was a housewife, mother of three, and president of the PTA at Woodrow Wilson School. She kept her home and her children clean, she prepared hearty meals that she believed to be healthful, she helped with homework and with yard work, she went to church on Sundays, and in her spare time she volunteered. She and others of that first generation of Heart Study participants no doubt saw themselves as typical Americans. It is only in retrospect that this generation, admiringly referred to by Tom Brokaw as "the Greatest Generation," has been revealed as people who accomplished amazing personal triumphs against enormous odds. Basic food and shelter were often hard to come by during their youth and formative years, education required tremendous sacrifice, and many lost time by serving in the armed forces. Yet all this was quite ordinary for their time. When Gilcin Meadors, Nell McKeever, and, later, Roy Dawber, William Kannel, William Castelli, and I needed help, Langley together with other community leaders and thousands of Framingham residents were there to step forward. Without the likes of Langley, Sullivan, and Galvani, people deeply entrenched in Framingham who had earned the trust and respect of their neighbors, the Heart Study surely would have failed.

Researchers at the start seemed to know intuitively that the Study, no matter how far across the nation and the world its ripples would reach, was a local matter. They assembled a committee of the town's doctors, the group whose trust and backing Dawber would need in early 1950 when he took the helm.

But key members of the community had a vision of how to shape the Study right from the beginning. They had deep roots in Framingham, and they had strong personal and professional stakes in seeing the town succeed and move forward. It was from this pool that Meadors and McKeever created the Heart Study's Executive Committee of Residents. Even before thousands of their neighbors agreed to contribute the medical details of their lives, members of this committee were convinced of the Study's potentially sweeping value. As average as they may have appeared on demographic scales, they were a unique and diverse group, often having in common little more than residence within the town limits.

Meadors and McKeever found Evelyn Langley in one of the first places they looked for a volunteer with leadership skills: in charge of a PTA meeting. The principal of the school, Dr. Mary Stapleton, recommended her to the Public Health Service. "I'm one of the charter members. I helped organize it," Langley says. "They called us leaders."

Langley spent her girlhood on a farm outside Framingham at a time when her immigrant Italian father believed that girls shouldn't go to college. "My father didn't believe in educating girls," she says, now in her late eighties and still angry about it. "I decided I was not going to marry any man who was going to tell me what to do."

She was a leader in part because of her roots in town, though her earliest years were spent in the rural surroundings. "I lived on a farm. My mother canned all the vegetables and fruits, and we kept them in the cellar. We had cows and hogs and chickens. I walked two miles to school, up and over a hill. What I remember about the Depression was that shoes had to wait." She and her mother made bread daily from flour they kept in a barrel in the cellar. They got fresh eggs from their chickens, made butter and cheese from the cow they milked, and every now and then slaughtered a pig.

She still enjoys a nearly full-time schedule of volunteer work. She was a perfect choice for the Heart Study: informed and persuasive. She remembers one neighbor who was particularly adamant about his privacy. He didn't want anyone from the government knowing about his health problems. Langley had quite a time with him. But she herself had already fully adopted the message of the Study. She gave him her word, a personal promise, that the results would always be kept confidential. He signed up. "Years later, they found a spot on his lung," she says. The report went to the man's personal physician, and the early, precancerous lesion was removed. "Now, every time he sees me, he says, 'I sure am glad you took the time to walk up to my door.'"

She spoke to a lot of immigrants, recently arrived from Ireland or Poland or Germany, and still having difficulty with the language. Words like "epidemiology" and "electrocardiogram" must have sounded doubly foreign. But many of them signed up. "Once they had one exam, everything was fine. They knew it would be a good thing," she says.

Beyond reassurances about privacy, beyond promising the benefit of free exams, she said things that, though proved correct during the past half century, must have sounded like hyperbole or science fiction.

"I would say to people, 'When you come right down to it, wouldn't you like to be part of something that will benefit the human race?'"

The commitment to the Heart Study has continued through Langley's bloodline. Her three children are in the Offspring Study. By the time the Third Generation Study began in 2002, many of Langley's eleven grandchildren had been asking for years for their turn to sign up. Membership in Framingham is an honored family tradition.

One of Langley's grandsons, Lou Merloni, has played for the Boston Red Sox. Because of his grandmother's involvement in the Study, he knows enough to say a difficult but determined "No, thank you," to some of his mother's rich dishes. "He really watches his diet. He said to me once when I put out a plate of ravioli, 'Mom, don't make that stuff anymore,'" says his mother, Sandra Merloni, who herself has difficulty practicing what the Heart Study preaches. She still loves ravioli and cappelletti—dough filled with bread crumbs, cheese, and eggs and cooked in soup—and misses the fried-pepper-and-egg sandwiches of her youth.

Almost since the Study's inception, ancillary studies have been included and participants have agreed, in addition to heart- and brain-imaging studies, to aid researchers involved in studies of osteoporosis, breast cancer, Alzheimer's disease, and arthritis, as well as hearing and vision disorders. Sandra Merloni had just completed sleeping with a monitor for a sleep study when she said, "Anything they want, I'm willing to do if it's going to help in some way—not only for me, but for my children and grandchildren."

In that vein, some volunteers have agreed to donate their brains to a brain bank, for postmortem study. The joke among seniors in town, many with blue-collar roots, is that when they die, they finally will have the brains to go to Harvard. Langley remains a dynamo. Her daughter has retired from driving a school bus, and teases her mother for still having to set her alarm and go to work every day. It's not work, Langley says. It's up to thirty hours of volunteering at the Callahan Senior Center's gift shop. She keeps the books, sends out checks, and waits on customers interested in the donated goods.

She also works with elderly women who live in nursing homes. "People get the wrong idea about the elderly. They think we're just couch potatoes. But elderly people want to do something. If they can't walk, they want to use their hands. So we give them yarn, and they make baby afghans. Last year, they made about nine hundred little afghans."

She has been a local television personality, hosting a show called *Elderviews* on a Framingham cable station. She reads, knits, and embroiders. She once was president of a local chapter of the American Association of Retired People, and was active in a group called Silver

Haired Legislators, working with state representatives and senators on elder issues.

But Langley's heart still lies with the Study. "When they call me up and tell me it's time to come in for an exam, I know I have that ritual to do," she says. She has made the trip to the clinic twenty-seven times so far. "I am trying to give back to the Heart Clinic [Study] what they have given me. I always feel as if I am part of something bigger than myself. It's not just for the people who live in this town. Many lives have been saved because of the Heart Study."

BEFORE THE WAR, Walter Sullivan was active in town meetings when the issue was zoning and the size of lots. After the war, when Shoppers World arrived, he remained active in discussions of revitalizing the downtown. By then, on his walks through the streets, there were a lot of familiar faces, as well as a growing number of newcomers he did not recognize.

He started out as a teacher at Framingham High School, where he met his wife of fifty-eight years, Katie, who died in 1999. Photographs of her—a slight Irish beauty with a charming smile—adorn his home. They grew up in Framingham, but didn't meet until they were both teachers at the high school. When they married, she had to quit her job. Those were the rules in 1941. "The man was the head of the house. The women who married could no longer continue teaching," he explains. It wasn't until years later that the rules changed, and Katie Sullivan returned to teaching French to elementary school children.

When the war ended, Sullivan returned to teaching and attended law school in the evenings. They had one child before the war, two more afterward; they had their own house with a mortgage interest rate of 4 percent.

In the summers he worked on mosquito control for the Board of Health. With Stuart Foster, a professor at Framingham State College, his job was to find locations where potential breeding might take place. Health officials were concerned because a local hospital, Cushing General, was treating injured servicemen, some of whom returned from the Pacific theater with malaria.

He remembered the earlier contribution to the study of tuberculosis by Framingham residents when his boss at the mosquito project, David Moxon, approached him. Moxon had heard Meadors's pitch about a new study of heart disease, and passed the information to Sullivan. "I didn't know what it was all about for a while, but my boss told me they were coming, and would I be willing to serve on the committee," Sullivan says. "I said, 'Well, okay.' And he said, Okay then, you're the chairman."

McKeever filled him in about the Framingham Heart Study's purpose. "As a committee, we didn't initiate anything. We just tried to do what they outlined in the plan," Sullivan says. "One of the primary things was to introduce the town to the project and get publicity in the local papers. We even had a little radio station. Our job was mostly to squelch any fears if there were any, and to give encouragement to the populace to participate. We were to say, 'If your name is chosen, would you please do your best to cooperate with the Study?' I guess there were some fears, like 'Am I going to be pawed over?' But surprisingly, there weren't many fears. We didn't have to pull any teeth to get people to join." When McKeever said that Framingham's population was a good cross section of the U.S. demographics, Sullivan believed it to be true.

He was a great choice. Sullivan knew Framingham in his bones. He understood it the way you do the place that molded your character, set your moral compass, nurtured your youth, provided a haven for your family and the means for your livelihood, and gave you reason for civic pride.

As a kid, he knew its neighborhoods, like Lokerville and Coburnville and Saxonville. He remembered that shortly before the Great Depression, the Irish kids and the Italian kids had unspoken things to fear from one another. They formed the Torrey Street Gang, the Lokerville Gang, and the Coburnville Gang, and they eyed each other suspiciously. "We'd just as soon sock the other guys in the nose as say hello," recalls Sullivan. Looking back, it all seems so innocent, groups of immigrants and first-generation Americans getting used to each other's ways as they all got used to America. And indeed, those bursts of rebellion had largely mellowed by the time most of the kids landed together at Framingham High School.

Sullivan grew up in a town small enough for everyone to recognize everyone else. In his childhood, he knew the peddlers who delivered cuts of meat, bottles of milk, fresh vegetables, and warm baked goods. When the icebox needed ice, his mother would put a sign in the window reading either *15 cents* or *25 cents*. "That's how the iceman knew how big a block we wanted. It was my job to empty the drip pan underneath." He even learned the names of the horses that pulled the delivery wagons. There were so few cars that after a heavy snow, a cop would close off Dennison Avenue to traffic. This would open the way for all the kids in town to spend the day sledding down the hill, provided they first piled sand at the bottom to prevent sledders from careening into the sporadic traffic on Route 126.

Life had a predictable regularity. Sunday was a big meal of roast beef or stew, Monday and Tuesday the Sullivans ate leftovers, Wednesday was meat loaf, Thursday was corned beef and cabbage, and Friday was fish. Saturday was beans that his mother started soaking on Friday. Like other working-class immigrants, the Sullivans grew up just north of the railroad tracks. New England families who claimed long ancestry in America owned the larger homes around the town square and farther to the north. Framingham Center was the preserve of the old Yankee elite. "We thought they were on a higher level than the rest of us," says Sullivan. "They probably thought so, too."

Sullivan passed the state bar examination in 1947, but continued to teach until his law practice took off. He was working full time as a lawyer soon after his first exam with the Heart Study. He had learned to smoke during the war when servicemen could get a carton of cigarettes for fifty cents. "Sometimes, when I was going to school, I smoked to stay awake." In the early 1960s, he remembers reading a *Reader's Digest* article about the dangers of smoking. If it was based on a result from the Framingham Heart Study, he didn't know it. All he knew was that it scared him into quitting. Study researchers "always asked if we smoked, and how many packs. They still ask."

By the time Katie and Walter Sullivan's children were in their thirties, they were ready to become volunteers in the Offspring Study. Their daughter, Marie Cox, lived overseas for many years, but has always managed to return for her exams. "Every time I would go to a doctor in France, or England, or Switzerland, the doctor was impressed that I was part of the Study," Marie says. "They all wanted to hear all about it. They knew the town of Framingham. The first time that happened it made me realize how important the Study was. People all over the world were watching it, paying attention to it." No one from the Study staff ever told her to eat differently, or exercise more. "But filling out that sheet made me think about what I was eating, how much I was exercising. I learned to cook differently, and our kids grew up eating differently. They're very aware of good nutrition and the importance of exercise," says Marie, who has once again settled in her home state, near Buzzards Bay.

VICTOR GALVANI served with Sullivan on the Executive Committee of Residents. Galvani's father was a laborer from Piacenza, a man who never learned to read or write, though he mastered arithmetic and eventually could sign his name. Labor was seasonal in Massachusetts, and Galvani's father and mother would return to Italy for work during the winters and springs of the first years of their immigration. It was for that reason that Victor was conceived in Framingham but born in the spring of 1914 in Piacenza. The family moved permanently back to Framingham when Victor was two. His hardworking father valued education for his children, and after high school, Galvani would hitchhike his way to Boston College and back. He remembers law school tuition at somewhere around $200; back then it was a lot to scrape together. He became a lawyer in 1937. While he was a student, he never ate in a restaurant, never bought a class ring, and never went to a prom. He was thirty-one before he owned his first car.

"The reason they threw me on the committee was that I was a selectman from 1948 to 1954," Galvani says. That made him a community leader.

"I'm a real nut on geography and population. In a period of about fifty years, Framingham's population was still 23,000. Then the housing boom started, and by the time I terminated my office term in 1954, we had 28,000. At that time, we had four major population groups: we had the English group, a large Canadian group, both French-Canadian and English-Canadian. Then we had the Italians and the fourth group was the Irish."

Galvani, now ninety, still goes to work every day to his law office, now run by his son, Paul. He vacations in Florida, and each year when he returns, he swears that he's going to stop working. "But then I have nowhere to go, so I end up coming back. I'm here every morning at 7:30. I putter around. I do wills or real estate. It's a young group of attorneys, and I enjoy being with them."

He believes he inherited his longevity genes from his father. His mother, who had diabetes, died of heart disease at the age of fifty-one. But his father lived to be eighty-six. "My eating habits have always been poor. I have an inordinate passion for Italian cold cuts and potato chips. I know I've broken all the rules. I've never given up my passion for salt. Now, my doctor says, 'Hell, if you reach 84, 85, 86, why worry about it?'" As with the rest of Americans, not all Framingham participants do what they know they ought to do. Some, like Galvani, get lucky.

Until recently he served on the board of the Friends of the Framingham Heart Study, a group of participants who decide how to spend money that is donated to the cause. When Study members die, their families often request donations to the Friends of the Framingham Heart Study in lieu of flowers. The money goes to finance travel to a conference for a research fellow or a computer for a secretary or some other need that isn't met in the budget.

"I've always loved the town. I may be the most parochial or provincial member of the study," says Galvani. "I've never gone anywhere other than Florida. I still go to the little Italian club and play cards at night. We play for a drink.

"We all take a lot of pride in the Study. It's as if we feel we've accomplished something. We enjoy reading about the effect it's had on our culture. It makes you proud."

THE STUDY'S planners gave great thought to how residents should be approached. Neighborhood committees were organized along school district lines. Professionals from newspapers, radio, and advertising created a speakers' bureau. They formed subcommittees for publicity; for industry, represented by the thirty-five plants in town; and for small-business owners.

None of it could have happened without the grassroots group that formed the Committee of Residents. From the start, the planners vested these ordinary citizens with great responsibility, and granted them deep respect. In a statement of the plans for the Study, organizers acknowledged the importance of the community volunteers:

> After all the other committees have functioned, to this group falls the tremendous responsibility of door-to-door interview and solicitation in the random sample study of a population. Realizing the shift in emphasis from persons examined on a volunteer basis to one in which people are being requested to be examined, they [the citizens of Framingham] have formulated policy, and to some extent procedure. The difference between a prospective examinee saying "Yes" and "No" depends on who approaches the individual, and how he is approached. The aim is to find the right person to approach each individual.[1]

Those who agreed to be part of the Study would answer questions about their eating, drinking, smoking, sleeping, and leisure habits. They would recall, as best they could, the habits of their parents and grandparents, and details about the causes of their ancestors' deaths. They'd remove their shoes and get measured and weighed. They'd have their blood pressure taken, their blood drawn, and their lung capacity measured. They'd stare unblinkingly into tiny flashlights as doctors watched their pupils react. They'd hold thermometers in their mouths. They'd sit still while someone put a stethoscope to their chest. They'd breathe deeply and hold their breath on command while doctors recorded the tiny and distinct sounds of their hearts. As technology advanced, they'd be asked for even more.

Marion Kittredge was a nurse when the Study started, and has never missed an examination appointment. "This old body has been around for a long time. I'm very privileged to be a part of the Study," she says.

Morris Shapiro returned for his twenty-seventh exam from his home in Falls Church, Virginia. He was an attorney and a trustee of Framingham Union Hospital in 1948. "Everything was normal on my first exam. But on a later occasion, I told them that when I went to the gym, I could do all the exercises, but when I went out on the track, I had discomfort in my calves and tightness in my chest. The doctor put a stethoscope on my legs. They told me I had atherosclerosis in my legs, plus angina. I followed through." At the time, there was no bypass surgery. As the medical world learned about risk factors, Shapiro changed his diet, exercised more. Eventually, he had bypass surgery. He credits the Study with prolonging his life, but it happened indirectly, just as it did for anyone who was not a Study participant. "They did no treatment. They would merely report the findings to your internist," he says.

These community leaders and the people they signed up would be watched to see who developed heart disease and who didn't. Those who survived knew that the deaths of some of their loved ones took on new meaning. Walter Sullivan's wife, Katie, who was a participant in the Study for fifty years, died of a cerebral hemorrhage on June 12, 1999, at the age of eighty-seven.

Evelyn Langley's husband died of a sudden, massive heart attack in 1978. Widowed for more than a quarter of a century, Langley takes medication to lower her cholesterol, but otherwise is in great shape. If her husband had lived longer, perhaps modern treatment might have saved him. "I do think if it were today, they would have been able to do other things," she says. But he went quickly and suddenly, at age sixty-five, eight days into his retirement.

Of the original group of Study participants, the youngest survivors are now in their mid-eighties, the oldest over a hundred. In 2003, Anna Skinner died of a stroke at the age of 105. According to her obituary, she was "proud" to have been the oldest surviving member of the original Heart Study volunteers. For half a century, she never missed a biennial examination. The last four were performed at the Caldwell Nursing Home in Ipswich, but when she lived in Framingham, she walked to her Heart Study appointments. She never learned to drive, and would often walk the two miles from her home to downtown Framingham for errands. Until shortly before she died, she continued doing exercises at the nursing home.

Skinner grew up on a farm and graduated from Framingham High School. She worked as a telephone operator and a housekeeper, and had a thirty-nine-year marriage to George Skinner, who worked for the Massachusetts State Police. He died in 1961. She never smoked or drank, and she shunned junk food. She tasted pizza for the first time at a nursing home party.

Her daughter, Ruth Bauer, says Anna Skinner never changed her lifestyle habits, and attributed her long life to good genes. Bauer, her husband Robert, and their children, continuing the tradition, are enrolled in the Offspring Study and the newly begun Third Generation Study.[2]

Some volunteers who have retired to other places have made the trip for exams from Alaska, Hawaii, Spain, Greece, Kenya, and Guam. When Rita Hengesch came back to Framingham from Sarasota, Florida, for her twenty-fifth biennial exam, she encountered a friend with whom she had grown up. For the few who cannot get out, researchers go to homes or nursing homes to perform the exams. Where once they submitted to an examination that took four hours, their recent tests have been scaled back to two hours. In their eighties, nineties, and beyond, many of them are frail and get tired just moving on and off the examining table.

Rita Duran, an original participant, is typical in her assessment that she has "contributed to science in a very small way." And, says Sandra Merloni, "I think it's amazing what the whole country has learned from this Study. It's not just benefiting Framingham or Massachusetts. It's benefiting the whole world, really."

SIX

The Launch of a Gold Standard

Roy Dawber was an insider, New England born and bred. He was a practicing physician. He spoke like a Kennedy. He grew up on the South Shore of Massachusetts, worked at the Veterans Administration Hospital in Brighton. He had impressions of Framingham that stemmed from childhood, knew it had a manufacturing base, including a new General Motors plant, knew that it was near but independent of Boston, knew that it had a diverse population.

He was a born leader, well connected through wartime experience to top researchers and power brokers in Washington, most notably Cassius Van Slyke, the head of the newly established National Heart Institute. Dawber was an internist with the U.S. Public Health Service. He liked working with patients, but was getting restless and wanted to combine his clinical skills with research. He resisted every effort of the service to promote him to an administrative position. Then one day, says Dawber, "Van Slyke called me and asked if I wanted to take over this thing."[1]

Dawber had heard about the "thing," the Framingham Heart Study. He was intrigued, but a little skeptical. He was privy to the public health rumor mill. Two obstacles seemed to be working against the Study. Its dual mission, observation as well as prevention, had everyone confused. And the doctors in town were growing reluctant in their support. They probably never fully trusted a bureaucrat like Meadors, and his chaotic leadership did nothing to assuage their fears. They still had doubts about letting a group of academics and government officials keep an eye on them.

Dawber had heard stories of discontent with the Framingham project. "I told Van Slyke that I would do it provided I could find some evidence that the doctors in the community would support it. I wanted to go out to Framingham to talk to the doctors. I had heard some reports on what was going on in Framingham that weren't too complimentary, that there were squabbles going on in town. I didn't want to get involved in some caper that was going to lose." The naïve notion that science could somehow detect heart disease when so little was known was endangering the entire project. Without the support of the town's doctors, the participants would soon lose faith. It was a daunting challenge to take on the design and implementation of a twenty-year study. There was no guarantee of success, or even gratification. He needed, and received, top-level assurance that the project itself, since its transferral from the Public Health Service to the National Heart Institute (NHI), was now in the hands of scientists and on a unified course.

"The Public Health Service had established a unit that was supposed to investigate ways of screening for heart disease. You had to wonder why they did that. Heart disease involved a number of factors. It wasn't likely you would find any simple test. That was one of the reasons that it wasn't getting anywhere," says Dawber. And the notion that the Study would implement measures to prevent heart disease was entirely out of the question. Van Slyke told him the Study's new mission was clear. It was to be a long-term, observational study with no distracting attempts— futile at the time—to prevent or treat heart disease. Efforts would focus on one goal: unraveling the core features that led to the disease. "This former half-baked thing was turned into, 'Let's do an epidemiological study,' " says Dawber.

"Many who were involved in health education annoyed me a little bit because they seemed to think that people were skeptical of doctors, that people did not like doctors. My feeling was quite the contrary. Namely, even though people did not like the medical profession, when it came to their doctor, their doctor was a great guy and they had complete confidence in what he was going to do."

So to get the confidence of citizens, he had to gain the trust of the town's increasingly skeptical physicians. It all happened within a few days in March 1950. Meadors had requested and received a transfer.[2] Dawber came to Framingham and started talking to local doctors. An early document from the Study suggested forming a medical committee as a way of making them feel like "co-owners of the program." The planners knew that few would volunteer for a study their personal physician did not approve, and Dawber began rebuilding support that had started to erode under Meadors. He walked a fine line, using his clinical expertise to gain trust; at the same time he continued to assure the doctors that he, and the federal government he represented, wouldn't snatch away their patients. "I went out to Framingham and looked up a number of physicians in town. I think the doctors accepted me not just as a public health guy but as a practicing physician. That was very important, that they be talking to me doctor to doctor, and not to a public health official. Unless the doctors in town are with you, it's a waste of time."

But while they liked Dawber's credentials as an internist and his history of taking care of patients, they didn't want him to take over the care of *their* patients.

It was a bad time for a federal study. "We tried to get various medical agencies to endorse what we were going to do in Framingham," he says. But try as they might with the Massachusetts Medical Society, which publishes the prestigious *New England Journal of Medicine,* the institution never gave them an outright endorsement. "The most they ever said was that they wouldn't oppose us," says Dawber. But, perhaps because he spoke their medical language with the proper New England accent, local physicians listened. As steadfast as they were about avoiding government intrusion into their work, they were also baffled by the epidemic of heart disease. Their patients had high blood pressure, too often escalating out of control and toward disability or death. Their patients were dying in their prime. Like doctors everywhere, they were frustrated by their inability to help.

"Fortunately, I got to know most of these guys in town, and I think they trusted me," Dawber recalls. "I said, 'I've got no ax to grind; I'm not involved in your treatment decisions. What we find, we'll give to you. You use it for what you want.' I wasn't doing any private practice. I wasn't competing with anybody. We weren't treating people; we were examining them. It was a matter of mutual trust. Finally, they agreed we were honest people not trying to promote socialized medicine in the United States."

Simultaneously, he worked for the trust of the citizens. He picked up on the goodwill established by Meadors, and added his own. He set a hands-on tone, as he, and every director and almost every physician-researcher who followed, took a fair turn in the rotation to personally examine the participants. No physician at the Study was simply an administrator or simply a researcher.

With the goodwill of the townspeople and the trust of the doctors, the Framingham Heart Study, tentative and conflicted for nearly two years, continued in quiet earnest.

Without guidelines and without precedent, Dawber had to design a study. He brought discipline, scientific methods, and rigor to a brand-new field of science. The NHI had developed a list of twenty-eight hypotheses, and those became the scaffolding upon which he would build the Study. In Bethesda, Felix Moore, the institute's new Chief of Biometrics, evaluated Framingham's statistical methods. He and Van Slyke visited Framingham, meeting with key people on the advisory committee to ease the transition to Dawber's directorship.[3] Moore's expertise was the foundation of the Study's rigorous design and statistical methods that would hold up for half a century, initially setting out to see if men were truly more vulnerable than women; if family history played a role in heart disease and hypertension; if ulcers or colitis—symptoms thought to be connected to stress disorders—increased risk; if smoking, drinking, or inadequate sleep had a role in the development of cardiovascular disease; and what roles obesity and high levels of total cholesterol played.

Under Moore's guidance with Meadors, then Dawber, the age of 30 to 59 was settled on for the Study group to avoid having too large a proportion of people with preexisting cardiovascular disease. Moore also predicted that following a population of approximately 6,000 people would ensure statistically reliable findings.[4]

Moore was a statistician reluctant to include elements such as psychological stress or occupation because they could not be adequately measured. He possessed what was then a rare talent for applied statistics and an appreciation for random sampling. According to Gerald M. Oppenheimer, "During Framingham's first year under the NHI, he was probably the principal architect of its scientific transformation."[5] And Dawber rejected questions about sexual dysfunction or psychological distress for fear of alienating the volunteers.[6]

Dawber's reputation as a clinician and administrator was solid. But he was not a cardiologist or epidemiologist. He needed a core team.

Patricia McNamara had been hired by Meadors the week before Dawber replaced him as her new boss. She stayed with the Study for more than thirty years as the data manager and analyst who computed and stored millions of health measures. Less than a year later, Dawber hired William Kannel, who would become the Study's third director in 1966. There were no rules to follow except those that Dawber hammered out: learning by doing.

"You see, we were trying to look at eighty variables in 5,209 people. We had no copy machines. We had to use carbon paper. No electronic calculators!" Kannel says. This was another era. "The [IBM] punch cards, they were the only technological advance that was there early on. This [IBM card-sorting] machine was like a big piano. You'd put all these punch cards in, and it would take eight hours to do what you could do in half a second now on a computer. Not only that, it would mangle a dozen or so cards every now and then. You'd have to stop everything and repunch them."

Even as late as 1984, when I began working full time at the Study, the names of the more than five thousand volunteers were kept on a rotating Rolodex file the size of a ship's wheel. When we wanted to look up an individual participant, we'd spin that huge wheel looking for a file card that would start the trail to locating volumes of information.

Dawber recalls the weight of responsibility. He knew exactly where the buck stopped. "When we started out, Van Slyke said if it's a success, you get the credit. If it fails, you get the blame," he says. "I always believed that probably something was going to come of this, but I was a little bit leery about how fast. I'm glad the data turned out as they did. Otherwise you never would have heard of the thing."

He was preoccupied with concerns about getting the Study off to a fresh start. He was working in a field where he knew gratification would come years, even decades, later. The observations made by the Study would be meaningful only if they were absolutely consistent over the natural life span of 5,209 human beings. He would need extraordinary patience.

He sought out the best minds of the time for advice and guidance. "We got this committee together, very informally," Dawber says. "It was intellectually challenging. Paul White was one. He was personally interested in this project. In fact, he probably was the one who suggested to Van Slyke that he should do this Study. Paul White would come out quite frequently to ask how things were going. His prestige was very valuable in this bit. It was also valuable in getting a full committee together. We had a committee of the top cardiologists in Boston, and we'd meet with them and discuss what ought to be looked at."

Even the top cardiologists were stumped, and their misconceptions were legendary. But they discussed every theory. One was that aspirin could reduce the risk of heart attack. The committee talked about including the amount of aspirin taken by Framingham citizens in the list of questions, but they decided, even with 5,209 subjects, the numbers wouldn't be large enough to form any conclusion about aspirin's benefits. It still makes Dawber regretful: "We could have had that [evidence] sooner."

They knew that patients with a rare form of extremely high cholesterol, called familial hypercholesterolemia, suffered early heart disease. "It was only beginning to creep into the thinking that cholesterol levels that were in the normal range might not be good for you," says Dawber. At the time, cholesterol levels as high as 300 were considered in the normal range. One of Framingham's major contributions was the revelation that levels of cholesterol, or blood pressure, or weight, could be average and still dangerous. Being average in an overfed, sedentary society can be lethal.

"When it came to weight," says Dawber, "we thought, 'Does it make much difference if you're overweight, if you're not really fat?' Most of the guys on the committee thought there probably was a difference. There was a general feeling that people who were overweight were more likely to get in trouble. But they were thinking of people who were very fat, not in the normal range. We didn't have any figures. We didn't have proof."

They talked about exercise. "For example, Sam Levine had been doing a study of Harvard athletes. He observed that during very strenuous exertion, some athletes' blood pressure rose tremendously. He predicted people who respond that way would get more heart disease," says Dawber.

They debated the relationship between diet and blood pressure.

They speculated about how many extra pounds were too many pounds, and how much fat in the diet was too much fat. They discussed smoking, even as some of them used the ashtrays. Kannel would often try to sit next to McNamara at planning meetings. "He knew I always had cigarettes. He was always bumming cigarettes," she remembers. The smoking camaraderie was short-lived. As soon as the data from the Framingham Study began to show that cigarettes increased the chances of heart disease, Dawber banned smoking at the Study site, and Kannel and McNamara quit not long after.

They pored over animal research results and human autopsy findings. They read the Metropolitan Life Insurance tables on smoking, weight, and causes of death. They packed their best guesses into the questions they would pursue. "These were all matters of belief or opinion without any evidence to back it up," says Dawber.

They began designing the elements of the Study using the beliefs and opinions of the most respected cardiology thinkers of the time. In doing so, they would create a vast amount of evidence that would serve to launch thousands of heart disease research projects around the world.

They already had data from the first round of exams, and Dawber reviewed every fiftieth record from these. He studied the handwriting, looking for Meadors's scrawl versus others' notes, seeking clues of consistency. He found variations, and concluded that there might have been some incorrect measures. So on April 28, 1950, there was a change in the way blood pressures were taken. Every switch in procedure since has been performed with an eye on maintaining continuity with previous procedures while incorporating improvements.

Dawber established a protocol for measuring blood pressure. It is the average of two or more measurements. It is never done in a rush. Participants are seated, at rest, for at least five minutes before it is measured. The blood pressure cuff size is standard and the arm is bare. A cuff too small could give a falsely elevated reading and a shirt sleeve could interfere. William Castelli, in his current role as director of a heart disease prevention program at MetroWest Medical Center, uses the Framingham technique. The patient sits for several minutes. Then Castelli might point to the pond outside the window over his shoulder, and say, "Look at the water. See yourself in a boat, fishing. Relax." It's an important way to reduce what's called white coat hypertension, in which a patient's blood pressure rises from the rush, anxiety, or fear of being in a doctor's office.

As a data manager, McNamara was a woman ahead of her time working in the largely male world of science. She says being a woman was never a problem, except for one niggling memory. The group had funding for an added position, and the physicians on staff were intent on hiring another doctor. McNamara said, "What are you going to do about typing? I don't type. I'm never going to type. I think you ought to have a secretary." The physician résumés were filed for future reference, and the Study got the secretary it needed.

Some of the source information they gathered from the early days is handwritten in penmanship that one simply doesn't see anymore. It is written to be precise, written so that a 4 could not possibly be confused with a 9, or a 6 confused with a G. The pages are the first draft of a treasure trove of data for posterity. Most of it was typed, on manual typewriters, with multiple carbon-paper copies when Wite-Out was still a technology of the future. In the columns, every comma and decimal point lines up. These are skills that once separated the executive secretaries from the secretarial pool.

But the paper is just a record. They were observing lives and deaths. Each day, a clerk from the Study would go to Framingham Union Hospital, having been granted permission to look over the admission roster. If anyone from the Study had been admitted, she would code the information on a punch card. Did the person have a heart attack? A stroke? An attack of angina, or chest pain?

As the years went by, Study staff had to verify deaths, largely centered in town. But some participants had begun to move away from Framingham. Clerks would scan newspapers, at first locally, then far and wide, for obituaries of participants. But an obit wasn't solid enough proof. Even a death certificate was insufficient evidence to assign a cause of death. Each passing had to be confirmed with a letter from the decedent's physician, hospital records, or a coroner's report, listing a cause. When these sources were unavailable, next-of-kin interviews were carried out in an effort to obtain information and pinpoint a cause. Only when all else failed would the investigators rely on the often inaccurate reporting on the death certificate.

Letters went out to city and state health departments around the country whenever someone in the Study died: "We believe a person whom we have been following in our Heart Disease Epidemiology Study has died within your jurisdiction, and we would appreciate getting a copy of the death certificate."[7] The number of deaths, of course, mounted with each successive year of follow-up. And each death represented both a loss and a gain. The loss was on the individual level, the inescapable ending of a life. It was also the loss of a volunteer to further observation. The gain, in the cold eye of science, was a critical clue, an event that would add to the store of data. When volunteers died, they left behind grieving loved ones, but also data on their habits and their health.

The Framingham data Evelyn Langley's husband left behind after his death in 1978 showed that, at five feet ten and 250 pounds for most of his adult life, he was obese. He had diabetes and he smoked. He loved American-Italian foods like pizza and lasagna with lots of cheese. Until the last few years of his life, he enjoyed a high-fat diet. Because of the Heart Study's published findings, to which he and his wife and neighbors contributed, his personal physician told him to quit smoking. He did. When he and his wife married, she outweighed him by five pounds, at 145. During the course of their marriage, his weight ballooned. "He had a very stressful job. When he came home, he would eat, have a bottle of beer, and fall asleep. The Heart Study made him decide that he should lose weight," said Evelyn Langley. He lost fifty pounds during the last year of his life, but the changes were too little, too late. He died right after a snowstorm. He had hired someone to shovel the driveway, and had walked through the drifts to move his car when he collapsed, a typical sudden death due to coronary heart disease.

The data from the participants' clinic visits were transferred to IBM cards. Each card had only eighty columns. It wasn't enough. Blood pressure alone was measured on every participant three times. Then it was coded with the number 0 if there was no indication of hypertension. It got a 1 if the systolic and diastolic pressures were both high on the same reading, a 2 if the systolic only was high, a 3 if the diastolic only was high. Then the card was coded an 8 if the systolic or diastolic pressure of each reading was high. The coding was 4 if all readings were borderline; a 5 if one reading was borderline, one abnormal, and none normal; a 6 if one reading was borderline, one normal, none abnormal; a 7 if one reading was normal and one abnormal; and a 9 if only one reading was available and that reading was borderline.

And that was just blood pressure. Dawber and his team had to design data coding so thorough it would hold up for decades.

The first exam questionnaires included age of parents at the time of their deaths; known familial history of cardiovascular disease; history of rheumatic fever, chorea, diabetes, heart disease; present acute conditions or outstanding complaints; complaints of pain; and drugs presently taken. Height, weight, temperature, respiratory rate, and mental state were recorded. The participants' hearts were X-rayed and listened to, and their electrical activity was traced on an electrocardiogram. Participants left behind samples of urine and blood. Doctors looked at their pupils and retinas, felt for lymph nodes and thyroid enlargement and liver tenderness. They pressed stethoscopes to chests and listened for abnormal breath sounds and heart gallops or murmurs. Nurses measured chest circumferences when the volunteers inhaled, again when they exhaled. They noted body types—thin, muscular, and obese.[8]

The staff knew that measurements had to be meticulously consistent. When volunteers were weighed, stocking or bare feet was the rule. Clothing was a skimpy johnny or light robe.

Every measurement, every sound, every answer, and every observation was assigned a code number. Three times a week, Dawber and McNamara, soon joined by Kannel, along with the young physicians who rotated in and out of the Study, would meet, compare notes, and ensure that they were all using the same codes to record the same observations. Then they combined measures, and coded the combinations. They would take the coded blood pressure variable and combine it with cardiac enlargement as seen on X-ray and electrocardiogram. Then they'd combine the definite hypertension code with a finding of heart enlargement on X-ray only, then with electrocardiogram only, to make a determination about the presence of hypertensive heart disease. And on and on with more combinations than they could possibly use.

A simple measure could be sliced in more ways than anyone was capable of analyzing. The doctors struggled to cross-classify the numbers in ways that held meaning. There was blood pressure measured supine and sitting, in smokers and nonsmokers, in subjects with and without diabetes. When they considered all the permutations and combinations, categories proliferated, and the number of events in any one category was too small to analyze. They needed a new system. Statisticians at the National Heart Institute responded by inventing something called multivariate analysis, a method of calculating the relative importance of each of several factors that appear to correlate with each other and with the occurrence of disease. It's a method that is the core of virtually every epidemiology study done today, where researchers, if they are to be taken seriously at all, must adjust their findings for age and gender and a host of other critical variables. By rescuing the Study from drowning in an ocean of incomprehensible numbers, scientists working with Framingham data came up with a way to add credibility and specificity not only to Heart Study findings, but to data from all studies.

In Framingham, the participants were identified only by number to protect their privacy. "I hired clerical workers from outside Framingham, to further ensure confidentiality," says McNamara.

The IBM card-sorting machine was a memorable piece of equipment. Huge and isolated in a small room, it made a noise that shook the walls and rafters. "It was about four feet long, a foot and a half deep. The racket was terrible. About five thousand cards might go through in ten to fifteen minutes. If I wanted to look at blood pressure and some other factor that had to do with whether you had an MI (myocardial infarction, or heart attack), I had to make up these new IBM cards to get all the data in," McNamara remembers. Then she'd have to endure pandemonium as the machine ran the cards. As she examined piles of statistics, the noise was sometimes beyond endurance. In one of the Study's many locations, she had to share an office with the infuriating machine.

The coding process bothered some of the young doctors. They were doctors, not coders, they argued. They were diagnosticians, not number punchers, and were unwilling to memorize the dozens of categorical codes that were unique to Framingham. They balked and rebelled, so Dawber and McNamara accommodated them by simply rewriting the forms, putting a number next to a trait or diagnosis. "When the examining physician circled 'yes,' there would be a number there, too. And that would be the code," McNamara said. It was the same information, the same process, but all the physicians in the clinic felt better about what they were doing.

The Study would move from one site to another over the next few decades. "We moved and moved and moved and moved," says McNamara, as though still weary at the thought of packing all those boxes full of data. "We started out in the corner building. Then the hospital built a place within the hospital for the clinic. But we were jammed in. I must have had an office as wide as from here to this window," she says, gesturing at a space of maybe ten feet by eight feet. "Dawber had a room down on the corner. No other doctor had an office. The doctors used the examining rooms for their offices when the exams weren't being done. And then next to my office was a little bit larger office. I had three girls in there doing clerical stuff."

Once, they landed in yet another Victorian house. Whether irony or poetry, it was a house formerly owned by Mary Baker Eddy, founder of Christian Science, the religion that opposes medical intervention in favor of prayers.[9] "There was this one enormous room on the side. In the back of this huge room were two fair-sized rooms. By then I think I had a staff of five people, all in this big front section."

The offices were always makeshift, always part of the nearby hospital's holdings. The first was in one of a half dozen or so abandoned frame houses the Study leased from the hospital, initially for $225 a month.[10] "The hospital would keep tearing them down and replacing them with parking lots. We were once in the hospital itself, but then they wanted a [psychiatric] unit, and they promptly booted us out. The places we were in were really shabby, and the hospital never wanted to fix anything because they knew they were destined for bulldozing. Once the ceiling fell on one worker's head," Kannel says, referring to Patricia McNamara. She wasn't hurt. "I think the hospital started feeling some legal overtones, so they sent someone over to fix things up. But we were always considered transient."

Paul Dudley White would stop by to advise and consult with researchers. White was a chronic optimist. Writing in 1951, he mused on the history of heart disease:

> The state of the world is not so bad as it is painted. Newspapers are filled with gloomy headlines, and it has become almost a fad to publicize the distressing events in current history and to foster fear and pessimism about the world of tomorrow. Even in medicine there is an overemphasis on the alarm reaction and the effect of the unhappy emotions on body, mind and soul. Worse events than present ones were taking place a generation ago in many different parts of the

world, but the press was less avid then. . . . What we did not know of we did not worry about.[11]

Having been one of the first to bring the electrocardiograph to clinical research, White kept abreast of technology. But his legacy, undoubtedly inspiring the Heart Study founders, was the principle that nothing took the place of a clinical record, and the patient interview stands above any technological innovation. "Listen to what the *patient* can tell you—it may be more important than anything else you do," he would say repeatedly.[12] Although he treated Dwight Eisenhower after Eisenhower's heart attack, most of his patients were quite ordinary. As late as 1963, he charged, for his unique specialty care, between $5 and $25 a visit.[13] "Paul White could have gotten any fee he wanted, but he never charged more than a modest fee. He was a guy who was interested in disease and not in making money," says Dawber.

This combination of goodwill, positive outlook, and professional competence helped boost morale at the Heart Study. White would look around at the makeshift clinics and offer solace. "He'd point out that you don't need elegant quarters to do good research," says Kannel. In the early 1990s the Study moved to a three-story brick building with concrete stairwells, no elevators, cramped offices, and all the glamour you'd expect from an abandoned convent. The building, once home to the Marion Sisters, retained a stark, nonmaterialist ambience. In 2002 the study relocated to its best site yet, a wing in the headquarters of a large international construction and consulting firm. We have a two-story atrium, and when participants arrive, the first thing they see is a large, newly designed logo: three interlocking, heart-shaped rings symbolizing the three generations of Framingham participants. We are still within a half mile of the original site.

But for years, researchers would occupy old homes that were scheduled for demolition. Hospital officials would upgrade antiquated wiring to temporarily accommodate the needs of the X-ray equipment and an electrocardiograph machine. Staff members would put up curtain dividers to separate office space from examination rooms. But they knew the land under the old houses was on the drawing boards for a parking lot, or a building, so they didn't want to put too much money into fixing up the place, only to have it knocked down. Tiny waiting rooms couldn't accommodate the crowd a heavy clinic schedule would create, but it seldom had to. One of the first rules of the Study was "Don't keep the volunteers waiting." "People visit, and they're always amazed," says Kannel of the many old quarters. "They say, 'You've done this work *here?*'"

Through the first years of the Study, the staff worked long hours in their makeshift accommodations. "We were concerned if we didn't work hours when the participants were free to come in, that they would not be allowed to take time out from work. So we worked evening hours. The housewives could come in during the day, but the men, we felt, would want to come in evenings. Our wives were not too happy, because we were working every weekend and evening," says Kannel. Dawber, whose wife was a nutritionist, says the whole thing wouldn't have worked for him without a cooperative family. "I had to run the clinic two or three nights a week, and Saturdays."

McNamara had fallen in love with the logic and certainty of mathematics at St. James High School under the tutelage of Sister Adele, a nun from Ireland. The Heart Study gave her an opportunity to apply with dedication what she knew about math to the field of medicine.

With patience, trust, respect, meticulous attention to detail, and a basic blueprint that has withstood the test of time, Dawber created a new way to do epidemiology and began gathering the evidence that would be used in thousands of studies and clinical trials. The field of cardiovascular disease prevention would advance, one step at a time. Indeed, advances would come so slowly, so steadily, that their individual importance was largely overlooked. When noticed at all, it was with a nonchalant question such as: Yes, but haven't we *always* known that?

While it often seems so, we haven't always known that high blood pressure, high blood cholesterol, smoking, a sedentary lifestyle, obesity, and diabetes are risk factors for heart disease. Some of the best guesses about the possible contributors to heart disease posited by Dawber and the experts of the early 1950s proved correct, others wrong, as they assembled evidence about its multiple causes. Today one cannot attend a major cardiology meeting without hearing pieces of that evidence cited. But well beyond medical research, the Study's findings have steadily changed the way Americans view their health.

SEVEN

Wresting Control from Fate:
Results That Changed a Culture

One day in March 1950, Norbert Renz came home early from the office. He told his wife, Anna, that he had indigestion, and he climbed the stairs to the bedroom to rest. Anna warmed some soup and brewed some tea. By the time she carried the tray upstairs, he was dead, suddenly and without warning, at the age of thirty-five. He left behind five children under the age of six. Brian, the youngest, was eight months old when his father died. For him, the story of his father's death comes through a filter of overheard whispers and family lore. As was typical of the times, Anna remarried and her children took their stepfather's name, King. Norbert Renz's death was like a bolt from the blue, but it was accepted as fate. No one really understood why a strong man— a former football player for Holy Cross, an officer with the Army Signal Corps during World War II, an enterprising businessman with so much love of life in him—would just lie down and die. But no one really talked about it. It was simply accepted as one of those mysterious tragedies that happen. Brian King grew up with a simple but overriding story in his head. One day his father "came home from work, went up to bed, and died," he says. The older Brian got, the younger the age of thirty-five seemed. He came to believe, though no autopsy was performed, that his father must have had a heart attack or a stroke.[1]

Kitty Walsh, Renz's sister-in-law, was a nurse in Brooklyn in 1950. She remembers Norbert's death, but, even with her training, could make no more sense of it than anyone else. "The doctors didn't know, either. They wanted a blow-by-blow description of what happened, and they said it must have been a heart attack," she says.[2]

After the war, Renz had settled with his growing family in Lowville, where he owned a franchise that distributed oil in the northern part of New York State. He didn't smoke, but he loved his meat and potatoes and, with the pressures of a large family and a business, he wasn't getting much exercise anymore. By the time Brian came along, the rambling, two-story Tudor with wraparound porch and bedrooms for five children was bustling and full. But Anna and Norbert missed their families in New York City, and they wanted their help and influence in raising the children. The day he died, he had just come back from his lawyer's office, where he spent the morning negotiating the sale of his business.

What Brian knows from deep within is that it's possible to miss someone you never knew. But quite apart from the emotional hole it carved in Brian's figurative heart, his father's death left behind a legacy of fear for his anatomical heart. Somewhere in the back of his mind for nearly four decades, he believed he was walking in his father's shadowy footsteps, on his own death march. "When I hit thirty-five, I'll tell you, I breathed such a huge sigh of relief," says King.

Half a century ago, Norbert Renz's passing was the sort of tragedy most families came to accept as something outside their control or understanding. It was the kind of death many coroners listed on certificates as apoplexy or collapse. Young men across the country were being decimated by heart disease. But scientists were not so accepting. Renz's sudden demise was but one loss in the growing epidemic of cardiovascular disease. Renz had no connection to the Heart Study, but his was one in a relentless death count that reinforced the urgency of its work. The reasons behind Renz's death were a baffling mystery, and the Study was still seven years away from issuing its initial major findings. By 1957, when the early results were published,[3] only thirty-four participants had developed heart attacks or symptoms of coronary disease.

Sudden death often was the first, last, and only symptom in those who died of heart disease. About a quarter of heart attacks weren't recognized as such by the victim or the physician, and a lot of doctors were still highly skeptical that there even was an epidemic of heart disease.

. . .

WILLIAM KANNEL WENT to Framingham in 1951, when working with Roy Dawber was the best of few job opportunities available from the Public Health Service. "The only alternative assignment I had was some outpatient department in Houston. So I decided to settle in here and give this a whirl," he says. The whirl turned into a lifetime of dedication. He would eventually follow Dawber as director, but he was a young physician just completing his medical internship when he signed on, and frankly puzzled that epidemiology could be applied to a non-infectious disease. The discipline had thus far been applied only to contagious illnesses like cholera and tuberculosis, diseases eventually tracked to a single infectious agent. He went on to publish more than six hundred papers from the Heart Study, including, in 1970, a landmark article assessing the role of blood pressure in stroke.[4] He helped to put Framingham on the map both in the public eye and in scientific stature. The phrase he and Dawber coined made little impression on him at the time. It was years later that a colleague, at a scientific conference in Mexico City, posed a riddle: Do you know where the term "risk factor" was first used in the medical literature? Kannel had totally forgotten that he coinvented the phrase.

Dawber, then most recently chief of medicine at the Brighton Marine Hospital in Boston, and Kannel enrolled in the Harvard School of Public Health's master's degree program in epidemiology, creating a blueprint for the discipline that others still follow. Because they were on the school's faculty in the Department of Preventive Medicine, they didn't have to pay tuition, an important consideration for both men, who were raising families. "We could only do it part time. We were busy designing the Study and setting up the forms," says Kannel. The courses they took in epidemiology and statistics overlapped with and informed what they were doing in their day jobs. They had wives and growing children, and were part of a generation of men who relied heavily on cooperative spouses to keep the home functioning. Sometimes Dawber and Kannel would speak to students about the Study and its unfolding results, anxious to get information out even before they were ready to publish. They were investigators willing to examine participants personally, to wait for decades for cardiac events to happen, and to review each record in great detail so that the information they gathered would be strong, verifiable, and pristine.

What they had as models in epidemiology was of little use for the work that lay ahead. Earlier investigators trying to work backward from patients with heart disease typically relied on general practitioners. They would send out questionnaires and ask physicians to fill them out based on the patients in their practice. "Well, those doctors had no vested interest in the study, and it's a pain in the you-know-what," says Kannel. "If you had people who have no vested interest in the results, they're going to resent being put upon. They'll see you as sitting up there in your academic ivory tower, writing nice papers while they think they'd done all the work." A dozen practicing physicians will have twelve different styles of taking and recording blood pressure. Some might round numbers up, others might round them down. Some might give patients time to relax. Others might casually slap on a cuff. In the parlance of science, the end results are not "crisp."

According to Henry Blackburn, a cardiovascular disease researcher who worked on diet and nutrition studies, Lewis Thomas, physician and essayist, once defined epidemiology as "thumbing through death certificates."[5] That about sums up the respect the field had among some prominent scientists at the time. The Heart Study investigators were determined to go well beyond that, and to get numbers that were precise from a study designed to give new prestige to their branch of research. Dawber began using the terms "little e epidemiology" and "big E Epidemiology" to differentiate the quality of results they would be getting. The Heart Study would be doing the latter.

The broad hypothesis was that cardiovascular disease was the result of a number of causes that work slowly and in concert to promote atherosclerosis. The resulting changes were minute and many couldn't be measured by the methods of the day. Typically, it was only after decades of continual progression of the invisible changes that symptoms developed and disease could be recognized. By then, it was usually too late. The Framingham scientists knew they were nowhere near wiping out heart disease. In fact, the first participants whom fate—and lifestyle, they would later learn—had marked for early onset of heart disease didn't survive. The Study couldn't help them. It could only hope that their deaths would help generations that followed.

Dawber and Kannel, with input from the National Heart Institute and leading experts including White, kept adding new hypotheses about the causal factors. For example, it was not until the fourth round of examinations, beginning in 1956, that a diet history was recorded.[6] Participants also reported on the number of hours they slept and their physical activity levels, at work and outside. A hypothesis was behind each test or question. No one knew what the result would be. Robert Sullivan, a second-generation volunteer, remembers "asking one doctor what they were going to do with all the information. He said, 'We're not exactly sure, but we suspect it's going to come in handy.' "

After the first round of exams, participants knew exactly what to expect. At the scheduled appointment hour, a nurse would escort the subject from the reception area to a dressing cubicle. That subject would move on to another cubicle for blood pressure, weight, and height measurements so the first cubicle would be free for the next new arrival. From cubicle to cubicle, for electrocardiograms, X-rays, and blood drawing, participants would pass each other in the hall in their robes and slippers, but they would not slow each other down. The staff and the subjects got to know each other professionally. Dawber instructed staff members not to talk with participants about their personal lives so as to keep exams moving on schedule and out of respect for their privacy.

In 1956, a group from the National Heart Institute came to Framingham and went through the same exam as the participants. One member of the delegation wrote:

> A favorable impression concerns the quality and *esprit de corps* of the staff in Framingham. One could not help but be impressed by the high caliber of the clinical staff with whom I met and the friendly, enthusiastic attitude of all working on the Study. These qualities undoubtedly have had a very tangible effect in assuring a high degree of cooperation on the part of the Study population. In the course of our discussions there were other indications that the residents and the medical profession of Framingham consider the Study an integral part of the community.[7]

Getting participants to come back year after year was a constant concern for Dawber and Kannel. They worked long hours to accommodate their participants. But mostly, people returned because they soon saw that the rules of the clinic put them first. Staff members were to be ready before the clinic opened. As time went by and participants grew familiar with the Study, they typically began to show up early. The respect was mutual. The early hopes of Evelyn Langley, for example, suggesting to participants that they could be part of something that would "benefit all of humanity," were coming true. The original group of volunteers would become the first study population examined to extinction. Their motivation is long past any personal benefit.

Not surprisingly, the most disliked part of the exam was having blood drawn. People hate needles. Doctors explained that the blood work was one of the most important parts of the Study. When a new element was introduced to the exam, doctors were expected to describe it thoroughly to participants before springing it on them. "I gave the staff a pep talk every week," Dawber remembers. "I said, 'I want these people to believe they're coming into the best private clinic in the country.' " The appreciation was passed from director to director. I learned it from Castelli, who learned it from Kannel, who got it from Dawber. Dawber's words, paraphrased, live on as a mantra: *These participants are the salt of the earth. They are doing us, and the world, a favor. They should be thanked at every opportunity.* I begin every clinic exam by expressing that gratitude and I've taught all of our young physicians to do the same.

But if the participants were satisfied, the researchers were getting restless. In the early 1950s, only four papers came out of Framingham, and they were primarily about the design of the Study and its procedures.

Kannel and Dawber believed, at least by the mid-fifties, that they could begin publishing some results, though the numbers of deaths and cardiovascular events were still relatively small. And they were hearing at least one significant voice of dissent from the National Heart Institute. Dr. Robert Berliner, assistant to director Van Slyke, was suggesting that the money spent on the Study would be better used for laboratory research into cures for heart disease. "Dr. Berliner thought clinical evaluations weren't going to prove anything. He thought the way to find answers was through laboratory studies, and that you wouldn't need a big number of people to find things out," says Dawber. Other voices said public education was meaningless, that it took years—decades, even—for everyday doctors to begin putting published findings into practice. But if Berliner was the Framingham naysayer, White was its guardian angel, urging patience for the data to unfold. His was an unwavering voice of support for the still unproved hypothesis behind the Study.

"I was getting impatient in the first few years, too," says Dawber. "It was a bit exasperating. You were just doing this job as a job. You didn't know whether you were doing anything worthwhile, if you were going to be able to make anything out of what you were getting." Kannel remembers being more optimistic: "We went along collecting data, though it was frustrating not to publish. This was a waiting game."

There are few "Eureka" moments in epidemiology. Alexander Fleming may have had such a defining moment when he discovered penicillin, or Isaac Newton when he formulated his apple-assisted theory of gravity. But in Framingham, as in epidemiology, scientific discovery is a jigsaw puzzle slowly pieced together. The conclusion is revealed long before the last piece is in place, even if some of its details are still missing. Years before the decisive evidence that smoking is bad or that exercise is good, Dawber and Kannel already had growing inklings of the result.

The slow pace of discovery made it a challenge for them to enlist new researchers into the Study, young scientists who were hungry to publish findings. Kannel remembers interviewing many of them, and watching their faces fall. "We had to tell them, we're going to keep examining these people, but we won't have results for maybe twenty years," he says. He would talk to friends in the profession, many of whom were lukewarm about the project. The word was that the Study was really just raking over information that had already been mined. And during the first years, only those directly involved in the day-to-day gathering of data had any idea of the potential importance of it all. Even researchers outside the inner circle of the Study had no way of knowing what was going on there, or what good any of it would do.

On one visit to the Study, Thomas Dublin of the National Heart Institute wrote: "I gained the impression that the professional staff is deeply disturbed and restive about the difficulties currently being encountered in the preparation of reports on their efforts. There is a feeling that the Framingham program is not sufficiently known in medical and scientific areas. The staff rightfully feels, I believe, that the preparation of reports from the Study has been unnecessarily slow and cumbersome."[8]

But inside that circle, researchers were kicking their smoking habits, starting exercise programs, and changing their diets. Dawber and Kannel were finding suggestive links between high blood pressure and heart disease, between tobacco use and heart disease, between elevated levels of blood cholesterol and heart disease. The findings forced them to examine their own lives, and they wanted everyone to have the information they had on all the things that put people at risk for heart disease.

But they couldn't tell the volunteers they were examining and reexamining every other year. The pact with the town physicians was that the Study wouldn't steal patients away. They could not, would not, proselytize. They wouldn't say to participants, "Quit smoking," or "Lose weight," or "Get off your duff and exercise." At times they wanted to shout it from the rooftops. "You'd find somebody who was a two-pack-a-day smoker, and we were beginning to see results showing smoking causes heart attack as well as cancer," says Kannel. "All we'd say is, 'See your physician.'" The investigators understood the medical chain of command. The only way to let the volunteers' physicians, and all physicians, know what advice to give was to publish the results of their years of observation.

Finally, the first paper of key findings, "Coronary Heart Disease in the Framingham Study," came out in the *American Journal of Public Health* in April 1957. Dawber and Kannel found, not surprisingly, that the incidence of heart disease was two to three times higher in men than in women. But they also saw a surprisingly high rate of angina among women aged thirty to sixty-two, a key discovery that exposed the falsity of the notion of heart disease as exclusively a men's problem. They also noted that in thirteen of forty-three men who died of heart disease, the first and only symptom was a coronary death. Almost half the men suffering heart attacks died, and one-third of them died suddenly. Before there was any medical agreement on acceptable levels of blood pressure, this Study reported that the higher the blood pressure, the greater the risk of heart disease, and that even those with borderline hypertension were at increased risk of disease. The same increased risk was seen with escalating cholesterol numbers. "Even with that small number [of events], it became apparent that these were not just chance occurrences but were related to these things called *risk factors*," Dawber says.

Another obstacle facing Dawber was official resistance from the National Heart Institute to disseminating research results to practicing physicians. He and Kannel had long been eager to share their findings, but differed with government officials on timing and tone. The conflict had been going on since shortly after the Study began, and would continue. Even as late as July 1968, a report from the National Heart Institute by a committee investigating the future potential of the Study was skeptical about informing physicians:

> Another drain upon the time of the Framingham staff has been its efforts at the education of physicians. Dr. Kannel and his colleagues have couched the analysis and presentation of much of their data in terms that they believe are understandable to practicing physicians. They have written several papers and given hundreds of talks to try to get their data promptly to practicing physicians. Serious doubt exists that this is a useful activity. Several studies have shown that the behavior in practice of physicians is not altered by presentation of data.

Cautious officials in Washington saw the data as preliminary and controversial. Kannel and Dawber regarded the results as profoundly important to the health of ordinary citizens. With those philosophical differences mounting, Dawber asked colleagues in research to help spread the word of what the Study had found. "The Framingham data were clear on blood cholesterol, blood pressure, and weight. There was nothing yet on cigarettes. When the data were published in 1957, I was working with animals mainly," recalls Jeremiah Stamler, who directed another long-term epidemiology study called the People's Gas Company Study. "I used to talk on the combined impact of cholesterol and blood pressure from my animal studies. Roy [Dawber] heard me talk. He said, 'The message needs to get out and we're forbidden from talking because we're employees of the federal government. We're under orders not to talk about any public health implications of these data.' He said they were told it was premature."[9] So while they were waiting for more data, Stamler and others did their best to discuss the new lessons about root causes of heart disease at medical conferences.

Six years after the first measures of total cholesterol were obtained on volunteers, the Heart Study reported that there was an association between cholesterol levels and coronary heart disease. It was a simple relationship—the higher the number, the greater the risk. And these results gradually were repeated in numerous other studies.

In 1961 Kannel and Dawber would publish a study that would be the beginning of a sea change in medical practice and in public attitudes about health and responsibility. The landmark paper was titled "Factors of Risk in the Development of Coronary Heart Disease—Six-Year Follow-Up Experience."[10] And inadvertently, in the simple words of that study title, they revolutionized twentieth-century medicine. They altered the slant of medicine from treatment toward prevention. They gave the public tools for living longer. Those title words would come to be inverted, changed in the common parlance from "factors of risk" to "risk factors."

But beyond the words, the concept that quite literally changed society was that multiple factors are involved in the complex pattern of risk that leads to heart disease. The Framingham Heart Study demonstrated that the disease is not random, but rather a predictable consequence of an identifiable pattern of exposure to risk factors. Understanding and applying the concept of risk factors is a critical first step in establishing a causal link between key predisposing traits and the occurrence of heart disease. The links create patterns that could be identified, quantified, and applied to people who hadn't yet developed disease. Though they had no symptoms, they had at last entered an era when simple measurements could identify those who were at risk.

And that concept gave physicians their first prevention tools. A risk factor, such as high blood pressure or a high cholesterol level, can be controlled. Understanding that there are common, easily measured, and controllable elements that lead to disease began to give people a feeling of self-determination.

Most people understand that while one risk factor in isolation typically does not spell doom, it is important to control those that are modifiable. The first paper to use the term pointed out that no single essential factor had previously been identified as leading to heart disease. "However, epidemiologic information has accumulated which now allows the physician to recognize certain characteristics of increased risk in patients he sees in his practice."[11] Two characteristics cited in the study leading to a disposition to coronary disease were hypertension and elevated levels of serum cholesterol.

Their published words, and hundreds of studies that followed, had a powerful effect. "Six years of follow-up experience in the longitudinal prospective study of coronary heart disease in Framingham have confirmed the widely recognized influence of hypertension and hypercholesterolemia on the development of coronary heart disease. These factors have been noted in clinical studies to occur in excess in persons with coronary heart disease, and in animal experiments to be associated with the development of atherosclerosis." But the big news was this:

> It is now demonstrated that these factors *precede* the development of overt CHD [coronary heart disease] in humans and are associated with increased risk of the development of CHD. . . . There can be no doubt that absence of these characteristics is distinctly advantageous since such persons demonstrate a relatively low risk of developing CHD. Whether or not the correction of these abnormalities once they are discovered will favorably alter the risk of development of disease, while reasonable to contemplate and perhaps attempt, remains to be demonstrated.[12]

That last sentence, in a nutshell, puts the role of the Framingham Heart Study in clear perspective. Medical science continually passes the baton of discovery from observation to laboratory studies to human clinical trials. Framingham had begun to lay the groundwork of observation on which scientists could build prevention and treatment strategies.

The popularization of the concept of risk factors has been life-altering for millions. It is a concept that is so simple, and yet it became the vehicle for translating medical research from the observations in Framingham into real changes in the way doctors practice medicine. We can quantify risk, look at its relations to individual factors, and design clinical trials to test whether risk factor treatment reduces cardiac disease. Before Framingham, researchers had some notion that there were multiple causes of heart disease. But, as in Cassius Van Slyke's 1949 mission statement for the Study, they used ivory tower phrases, like "etiological factors." Such a scientific term could not possibly engage the public, much less cause people to contemplate changing their behavior. Proof of the concept in Framingham, along with new packaging in the form of the term "risk factor," changed everything.

The understanding that people can actually *do* something about their risk factors for heart disease has led to a populace with better management of its health. "It's had a major effect on the way doctors practice medicine. It has provided a sense of urgency with regard to what doctors were formerly considering medical trivia," according to Kannel. In the middle of the twentieth century, doctors were talking about hypertension as a benign and essential part of aging. They even called it "essential hypertension."

> We said [Kannel explains], "Hey, it's not benign. It's not essential." We were telling them that a little cholesterol elevation doesn't cause any symptoms, but it can kill you. We said smoking is not something you can tolerate. Unrestrained weight gain is not a good idea. Regular exercise is.
>
> Doctors were dealing with completed catastrophes, and we were convincing them that a heart attack or a stroke in their patients should be regarded as a medical failure—not the first indication for [starting] treatment. This concept of preventive cardiology, that physicians can do a lot to protect their patients, is a major thing [the Framingham Heart Study] has contributed.

Coming at a time when little could be done to treat those with established disease, the elucidation of risk factors represented a quantum leap. With this information, the way was opened for an entirely new approach. Doctors could suggest early interventions such as weight reduction and salt restriction to prevent blood pressure in the high normal range from progressing to hypertension.

Reporters in the lay media were interested in writing about the Framingham Study, since it touched on so many basics of people's lives. Dawber told his staff in a memo, "Publicity is valuable, and local publicity necessary—nothing dramatic but pointing to what we are doing. We are running a different kind of operation than anyone else; most studies are not done as extensively as here and one of the chief factors is to convey the interest we have in the Study to the people who are co-operating with us. Publicity is necessary to keep interest in the Study alive, and to answer the question, what are you turning up? We have a moral obligation to the people who are cooperating with us in this Study."[13] He wanted the Study to be more in the public eye. Just as he and the other insiders were learning to stop smoking, to eat a lower-fat diet, and to exercise, they hoped that Americans would grasp enough to make those choices as well.

The term "risk factor" has taken on a life of its own and has lent itself to a way of viewing disease, recreation, chemical toxins, and the way we drive our cars. There is health-enhancing care and there are health-eroding risks we can take. By the 1970s, Kannel and Dawber realized the impact of their unintended mots justes. And Kannel, with the benefit of hindsight, sees that every new reported risk factor followed a predictable course. "Let me tell you the natural history of a new risk factor," he says. "When it's first discovered, the scientific community says, 'It's probably not true.' Then, years pass, and it's found to be true. Then the scientific community says, 'Okay, it's true, but it's not important.' Then more years pass, and finally they say, 'It's true, and it's important—but it's not *new*.' Every time we asserted something, it was first said to be bunk."

Inside science, there were skeptical peers. Outside science, there were powerful foes. Published findings from the Study angered not only the tobacco industry but also the egg, dairy, and beef sectors.

After the first decade, however, incoming results were strong and irrefutable. "We were not backed [financially] by anybody but the government, so we didn't have an ax to grind. We were free to publish whatever we came up with. Whatever it showed, it showed. I guess I never realized it was going to change the whole world—in a way," says Patricia McNamara. Physicians had been busy dealing with catastrophe, and the Heart Study was giving them a way to prevent it.

Kannel came up with a clever way of getting the word out on what was being found in Framingham. He was ahead of his time as a spin master, with an instinctive understanding of the competitive world of journalism. When a significant finding was going to be released at a national scientific meeting, he'd call the *Middlesex News* and give it advance notice. The local reporters loved getting a scoop, beating the bigger papers. The men and women in Framingham could be the first to read in their hometown paper that their Heart Study exams had led to the discovery that coffee was probably okay, or that cigarettes were definitely not, or that high blood pressure should not be left untreated, or that the average weight in a well-fed, sedentary society was too high. Later, they'd see the results on the national news.

Framingham was becoming famous. When Morris Shapiro took his family to the Grand Canyon in 1961, his young daughter developed an infection in her toe. They stopped in a small-town hospital in the middle of the Arizona desert. He told the nurse they were from Framingham, Massachusetts, and she said, "Oh, the Heart Study!" When Karen LaChance was on a cable car at Chamonix, France, a German tourist knew of her hometown from its scientific contributions.

It was a complex concept that Dawber, Kannel, and others took on in the Framingham Heart Study. With time, their patience and perseverance would pay off in the 1950s and 1960s with a discovery of identifiable risk factors for heart disease. The bonus for everyone alive today is that most of these risk factors can be controlled by eating healthy foods, exercising, and watching blood pressure and cholesterol level and seeking treatment if either is high.

Brian King, who has never been to Framingham, is in his fifties now. He walks for exercise, doesn't smoke, and knows he should lose a few pounds. He is aware of his blood pressure and cholesterol level. In other words, he realizes he can control his risk factors, and does his best to do so. The fate of his father, Norbert Renz, does not have to be his. Framingham has enabled him to reduce his odds of developing heart disease. For millions like Brian King, a family history is no longer destiny.

EIGHT

A Near-Death Experience

For the Framingham Heart Study, 1968 was a year of reckoning. Across America, it was also a year of assassinations, riots, and the struggle for civil rights. Abroad, it was the year of the Tet Offensive, a massive buildup of American troops in Vietnam, the capture of the USS *Pueblo* by North Korea, and Soviet tanks in Prague. With so much chaos and turmoil around the world and on the minds of Americans, the news of plans to terminate a study of heart disease did not at first cause much commotion.

The Heart Study was not then, as it is today, a widely recognized national treasure. The *Texas Heart Institute Journal* called it, in 2002, "one of the most impressive medical works in the 20th century. . . . Not only was the Framingham Study a milestone in the history of cardiology, but it has served as the model for many other longitudinal cohort studies. We remain indebted to those who initiated the study and thus became pioneers in preventive cardiology."[1]

But in 1968, one of those pioneers, Dawber, was not accepting plaudits. He was fighting for the Study's life, making phone calls, traveling across the country, determined to raise private money to keep it going.

Behind the scenes, Kannel and Dawber were worried that their life's work was about to come to a premature end. Today, they talk about the era without anger and with the pride of those who fought a tough battle and came out on the winning side. I learned about those difficulties for the first time when I read in 1982 a reference in Chapter Two of Dawber's book *The Framingham Study.* "[A]fter 20 years of operation, a review committee recommended that clinical examinations be discontinued, inasmuch as the major hypotheses had been adequately tested. This decision was not approved by many persons, both inside and outside the National Institutes of Health. The Heart Institute indicated that it would follow the recommendations of the review committee. Some who disagreed tried to have the decision reversed. This was not successful. . . ."[2]

It is a brief and understated description of several years of ambiguity. As public as the near-shutdown was, with newspaper reports and entries in the *Congressional Record*, I learned early on from Bill Castelli not to bring it up publicly. It would be regarded by some as a blemish on an otherwise illustrious history of the National Institutes of Health. But it is an important story to tell, illustrating how the devotion of the Study's founders knew no bounds, the loyalty of its volunteers was steadfast, and the value of its findings, two decades after a shaky start, was universally recognized.

By the end of the 1960s, the Study was facing stiffer criticism than in 1948. How much more could be learned from an investigation that already had fulfilled its twenty-year mission? Would research money be better spent on laboratory science? By the mid-1960s, those questions were being asked by scientists and physicians who had a say in the continuation of funding.

At the time, there were still those in science who believed that answers to disease would come only from beakers and petri dishes, that working with a human population took too long. "There was a group of people, you might call them purists, who thought science would only advance by working in the laboratory," says Dawber. Dr. Theodore Cooper, then head of the National Heart Institute, probably was not among the doubters, Dawber believes. But he had his superiors at the NIH to contend with. "He was under constant pressure to close us down," says Dawber. And he was new to the Heart Institute, a cardiac surgeon with an interest in the possibility of creating an artificial heart. "He probably felt he should be implementing his own programs rather than simply extending programs that were developed before him," Kannel speculates.

In the face of uncertainty, a directive from the federal Office of Management and Budget instructed the NIH to cut personnel. Each institute had to absorb a share of the cuts. "Pressure came to have most of these cuts come from research programs outside of NIH proper," says

William Zukel, who was then associate director for epidemiology and biometry at the Heart Institute. Extramural field studies that various institutes were supporting were the first projects to be sacrificed.[3]

The Framingham Heart Study was vulnerable. It was originally envisioned as a twenty-year investigation and, without a compelling reason to continue, its time was up. It was located far from the NIH campus in Bethesda. In one motion, twenty-seven positions could be cut, twenty-seven slots freed from funding, twenty-seven salaries funneled to other intramural research programs.

There was also the problem Dawber and Kannel had anticipated years before. They were accumulating far too much data to analyze with available statistical methods. Manning Feinleib, now professor of epidemiology at the Johns Hopkins School of Hygiene and Public Health, began to work at the National Heart Institute in 1966 and eventually became head of the Field Epidemiology Branch that oversaw Framingham. He recalls the problem of what to do with the accumulated data:

> The key way of analyzing the data had been to look at every two-year exam by itself. Around 1966, the NIH became concerned about what to do with the Study. They were not enamored of it. Their interests were first laboratory studies, then clinical trials, and third epidemiology. So here was Framingham going along publishing these papers. Bill Kannel was a prolific author. We used to say that every time he took a plane trip, he wrote a new paper. But from the perspective of NIH, all of those papers were the same. There were the six-year results, the eight-year results, the ten-year results. By about the mid-sixties, they felt that everything Framingham could do had been done. All they would do is tack on a few more cases. It was a major statistical problem, how to look at the total data.[4]

In a few years, the development of multivariate statistical analysis would provide scientists with new and precise ways to look at Heart Study data, but it wasn't yet available when the Study was struggling to save itself. Without that method of calculating the relative importance of each risk factor and its relationship with other risk factors as well as with disease, researchers were in danger of becoming overwhelmed by useless data. In the beginning, optimism led Dawber and Kannel to take dozens of measures from participants and record and file them, often without knowing what use the information would have. Multivariate analysis would breathe new life into the stacks of data, but the two Study pioneers fought for survival before that salvation would come. What they did realize was that a gold mine of information still lay untapped and they needed more time to get to it.

The greatest pressure to close the Study was economic, with federal budget cuts affecting all departments. The Study had begun on a wave of fear of the epidemic of heart disease. But that threat, while still very real, was superseded by a new wave sweeping the country—the fear of cancer. By the mid-1960s, the focus of medical science was largely on unraveling its mystery, exemplified in President Richard Nixon's call to arms. Many politicians and bureaucrats wanted to free up the money spent on Framingham and put it elsewhere. The overriding question in Washington was this: Has the Framingham Study outlived its usefulness?

When Dawber realized that support was slipping, he resigned and took a position at Boston University, where he could raise funds for the Study. He made it his mission to save what he had created, and he knew he had to do it from the outside. He passed oversight of the Study to Kannel.

Kannel dealt with NIH bureaucrats and an onslaught of reviews of the Study's continued value, or lack of it. In 1948, the original planners figured two decades would be enough time to observe the lives of the five thousand men and women aged 30 to 60 and follow them as they reached 50 to 80, developed heart disease, or died. Some wondered if additional follow-up would provide new ideas and novel insight, and stimulate innovative approaches for the prevention of heart disease. Kannel was among the fiercest believers that more could be gleaned from the Study population, with its rich, cooperative source of medical information. Dawber, Kannel, and Castelli—architects, scientists, and ardent promoters of the Study from the inside—and Feinleib, Zukel, and others from the National Heart Institute felt there was much left to discover.

No other study in the country was in a position to monitor the health of an entire community with such precision and detail. If Framingham closed, it could never be replaced. The crisis happened just as researchers were on the verge of being able to mine another lode of data. By its very design, the Study's particapants were approaching ages when they would begin to encounter escalating rates of disease and death.

But, as always, Framingham lacked something the late 1960s was full of: glamour and flamboyance. The end of that decade saw the dawn of technological wonders that put a man on the moon. In future years, advances would bring computer technology that could address and solve complex problems in milliseconds, uncover cellular mysteries, and map the human genome. Cancer research, technological advances, and basic laboratory science had grabbed the attention of researchers. The war in Vietnam, as well as the national commitment to the space program, drained the federal budget and forced cutbacks in medical research. Epidemiology could not compete for headlines—or for research dollars.

"It was a real crisis. There were these bench science types that tried to torpedo Framingham. They were going to it shut down," says Kannel.

A review of the ongoing value of the Study began in 1966, when a committee of the National Heart Institute was charged with evaluating all of the institute's epidemiology studies. The NHI sent investigators for site reviews, including William J. Zukel; Harold Kahn, senior statistician; and two other statisticians, Jeanne Truet and Dewey Shurtleff. A scientific review committee included George Burch, Alan Gittelsohn, George James, and Abraham M. Lilienfeld.[5]

Their mission was to analyze the results of the Study to date. Had it been scientifically productive? Had it met the original objectives? If not, how much more time would be needed? What were its most stellar accomplishments? Committee members were also charged with projecting the Study's value into the future. What were the areas of greatest potential? Did the Study need a change of research direction or of objectives?

In other words, what was the future role of the Study in research, training, and professional education? They looked at all of this in light of the relative status of epidemiological research compared with bench science and human trials.

The site review report was signed in 1966 by Abraham Lilienfeld, a Johns Hopkins professor and a big *E* Epidemiologist. His textbooks, *Foundations of Epidemiology, Preventive Medicine,* and *Reviews of Cancer Epidemiology,* are classics. He was also a pioneer in developing epidemiological methods for the study of chronic diseases, including cardiovascular disease. He understood the value of the Framingham Heart Study and was an avid supporter of its continuation.

By then, the clinic was in cramped quarters in Framingham Union

Hospital, and the annual budget was $336,000. The committee ac-
knowledged that the Study had accomplished much of what it had set
out to do.

"The prospective study has proved to be feasible. It has, in fact,
demonstrated that it is possible to study heart disease in this manner, and
it has served as a model and training-ground for similar studies in the
United States and elsewhere. Secondly, it has yielded data of considerable
medical and scientific interest," according to the committee's report.[6]

Published reports from the Study, at first slow to trickle out, were
appearing in large numbers, giving Framingham credibility and visi-
bility. Researchers had determined the incidence of heart attack, angina,
and death from heart disease in the general population—though not for
those older than seventy-five.

Physicians were gratified to have, at last, medically sound advice to
give their patients. Bench scientists were using the findings to develop
new theories and to begin to invent and test new drug therapies. Findings on cholesterol, blood pressure, and cigarette smoking stimulated
and focused basic research.

Committee members gave Framingham credit for allowing medicine
to identify men and women susceptible to heart disease *before* they
developed it. And in keeping with the baton-passing metaphor of science, this created a surge of interest, they wrote, in developing prevention strategies and treatment techniques.

The accolades in the report were hardly comforting. They showed
only that the original mission had been accomplished. But the Study's
staff believed there was much more to accomplish. Cerebrovascular disease, or stroke, had only recently been added to the Study's concerns.
Researchers would soon assess cardiovascular disease risk in relation to
HDL cholesterol level, coffee drinking, smoking, physical activity, alcohol, dietary habits, and weight change.

Dawber and Kannel were heartened by the committee's conclusion:
"It is difficult to be anything but enthusiastic about this study. It has
been a model of cardiovascular research of this kind. It has produced
data of great importance and interest and it has profoundly influenced
research and clinical attitudes. What it has produced could not have
been obtained in any other way."[7] Committee members recommended
a quick decision. Like the Framingham staff, they saw the potential for a

sustained flow of results, and backed the continuation of the Study for ten more years.

But the issue was far from settled. Senior officials and government bureaucrats had yet to weigh in. The Study's value may have been appreciated, but was it economically competitive with other research in which the country could invest? The situation continued to brew over the next four years. Little wonder that the Framingham staff, according to the committee report, "have experienced a loss of morale because of indecision regarding the future of the project."

"It was a big circus," says Castelli. Suddenly, it seemed all eyes at NIH were focused on Framingham.

It was an emotionally charged time among Framingham researchers, and there were clashes between the independent-minded staff in Massachusetts and the bureaucrats in Washington. Minor irritations and personality differences began to grate. The difficulties seemed to center on management style, not scientific substance. Dawber and Kannel were extraordinarily self-reliant and held the reins of the Study as though they owned it. Suggestions from Washington to follow basic rules of government made Kannel balk. It was a classic clash between federal policy makers and a group of devoted pioneers who had grown accustomed to a good deal of freedom in the start-up years.

Dawber had been cautioned, twenty years earlier, to guard against control of the Study being taken from his hands. He spoke of the warning at the Framingham fortieth anniversary in a speech at Boston University.

> Dr. Van Slyke was enthusiastic about the Study but also was realistic as to its possible success or failure. He assured me he would give his full support in terms of financial backing from the NHI and that neither he nor the staff in Bethesda, Maryland, would interfere in my conduct of the Study. However, he cautioned me that I should be prepared for future changes in the administration of the NHI, and that if we were successful in determining any important factors in the development of coronary heart disease, some future NHI administration might well wish to become more closely involved in the direction of the Study.[8]

The day of increased federal involvement was at hand, after Dawber and Kannel had had a long run of creative independence. They wanted the federal funding and the support to keep the Study afloat, but they bristled at the loss of their autonomy. Decades after the crisis, there is appreciation for the stability and devotion passed on in the succession of directors. "I think the key to this Study is the dedication of Roy Dawber, Bill Kannel, and Bill Castelli. They dedicated their whole lives to this project," says Feinleib. At the time, however, Dawber and Kannel had to be fearless—even obstinate. Kannel especially seems to have been unaware of his tendency to ruffle feathers. The mitigating factor that saved him, and helped protect the Study, was that he was an extraordinary researcher. During this period of turmoil, knowing that it was crucial to produce or face extinction, he published forty-eight papers in peer-reviewed journals. Whatever else he may have projected within the National Heart Institute, he redeemed himself with prolific writing. It was a case of his work trumping all else.

Despite the recommendation of the review committee to continue the Study until at least 1976, the National Heart Institute was not convinced. A more detailed follow-up review of Framingham was ordered, and Lilienfeld continued as lead scientific adviser. Kannel was worried. The committee issued several reports over the next three years. The competing pressures to spend the money elsewhere continued. He took the threat of closure as a challenge, and was more determined than ever to prove the Heart Study's continuing value. "We were just trying to hang on, but I made a conscious decision to demonstrate the importance of these data," says Kannel. "I decided we couldn't rest on our laurels; there was plenty of new information to be gleaned."

As uncertainty and controversy swirled around him, Kannel kept publishing. It was a classic example of an irresistible force meeting an immovable object. He could not be swayed from his commitment. His devotion to the Study had proved to be a double-edged sword. His tenacity was an asset in earlier years, when a less determined researcher would have abandoned the work in favor of projects offering instant rewards. But by now, Kannel's obstinacy was a source of irritation to a number of people in Washington. The National Institutes of Health's annual budget was rapidly approaching a billion dollars, and the growing rules and paperwork of government were encroaching on Framingham's independence.

Clearly by 1965, the Heart Study was already seen as a model of how epidemiological field research should be done. Hindsight shows us the additional value of the data from the three decades that followed. But foreseeing it was another story.

The most powerful argument in the Study's favor was that its participants were just approaching the peak age for the incidence of heart attack, sudden death, stroke, congestive heart failure, and peripheral vascular disease. By exam number eight, sixteen years into the Study, 531 of the participants had died and others, of course, could be expected to do so at escalating rates each successive year.

But Cassius Van Slyke, an early supporter, was dead. Robert Berliner, Van Slyke's assistant when the Study began, was arguing against continuation. "As we got near the end of twenty years, Berliner was still there. Van Slyke was gone. I'm sure [Berliner] was instrumental in encouraging its close so he could use the money for something else," says Dawber.

Lilienfeld's committee again considered several options that included stopping the Study, extending it indefinitely, continuing it for at least another decade, and involving other institutes of the NIH in order to thoroughly study the aging process. Once again, the committee was set to recommend continuation.

Then it was thwarted. The entire review process was ignored. Just before the committee's final report was due in June 1969, a preemptive directive came down on May 27 from NIH director Robert Marston's office to eliminate the Framingham Heart Study. Committee members were astonished and insulted. When Manning Feinleib saw the words written in response to the directive, more than two decades later, he turned somber and said, "Ah, yes. The infamous memo."

May 28, 1969
To: Director Theodore Cooper, NHI
[From:] William J. Zukel, Associate Director for Epidemiology & Biometry, NHI
Re: Directive from NIH to Eliminate Framingham Study

I wish to register a strong protest to the directive in the memo of May 27, 1969 from the deputy director, NIH, which specifies elimination of the Framingham Study by June 30, 1970.
I am astounded that the NIH is undertaking program decisions that should be the prerogative of the [National Heart] Institute and

secondly, that this action is taken before the presentation of recommendations on Framingham requested from the Epidemiology and Biometry Advisory Committee by the Director of the Heart Institute. A detailed review of the Framingham Study has been conducted by a project site visit May 13 and by full discussion of the Epidemiology and Biometry Advisory Committee on May 24.

To find that these recommendations which would be based upon an informed review have been superseded by administrative judgments in Building 1 [headquarters of NIH] casts a sad commentary on the use of scientific advisors of eminence and methods of administrative judgments affecting program priorities.

The recommendations on the Framingham Study will be provided by June 16 from the Advisory Committee as you requested. However, I believe that irreparable damage has already resulted from this NIH action and these will have broader implications than the scientific merits of the Framingham Program itself.[9]

Zukel's astonishment that the NIH was making decisions that should have been made by the director of the National Heart Institute is a clear indication that there was an intense internal battle between agencies. Despite the fact that a decision apparently had already been made, the advisory committee filed its report on June 16. Its summary, signed by William Zukel, read:

There is a clear national value of the unique resource provided by the Framingham Study, and the type of knowledge it is producing will not be forthcoming from any other population study. The medical and social importance of the cardiovascular diseases, cerebrovascular disease and emphysema in the age group reached by the Framingham population warrants active research attention within the NHI even if the opportunities for research on other common non-cardiovascular disorders of aging cannot be undertaken.

If the Institute is not allowed to make its own determination on how to allocate reductions of positions, the Framingham Program should be allowed to compete for NIH or Regional Medical Program grants or contract funds under sponsorship of one of the Boston Medical Centers.[10]

But Zukel's words changed nothing. In October 1969, Theodore Cooper officially announced that the epidemiological study in Framingham would be phased out by July 1970. The reason given was the lack of about $400,000 in annual funding.[11]

NIH slashed Framingham staffing and funding to skeletal levels. The plan was to pare back to five employees: Kannel, Castelli, and McNamara, along with statistical assistant Dorothy Costello and follow-up nurse Carleen Simoneau. Their job would be to analyze data from the final exams, and their cumulative salaries would amount to $64,000, with another $20,000 budgeted for travel, microfilming, and support services.[12] Their mission was clear: wind down the operation. There were no funds or staff to continue examining Study participants. Kannel began drawing up plans for a final questionnaire for the volunteers— this one a mere paperwork mailing, or as Dawber might have put it, "little *e* epidemiology."

It was just what Dawber had feared. But he had been working on a contingency plan. Zukel's final comment opened the door for Boston University, now with Dawber on board, to get involved. By the time the closure was announced, he was mastering the art of fund-raising for the Study from his outpost at the university. It was a career twist like nothing he had undertaken before. But, just as not knowing how to begin an epidemiological study of a chronic disease didn't stop him from starting one, not knowing how to raise funds didn't deter him from doing it.

David Rutstein, instrumental in launching the Study and choosing Framingham as the site, got involved again. He urged Dawber to raise money to keep it afloat. "So I got the [Boston University] medical school to endorse my trying to raise money privately to keep Framingham going. The [staff] fund-raiser for the school helped me do it," says Dawber.

Dawber approached insurance companies, drug companies—even the tobacco industry, which at one time pledged $250,000 a year for up to ten years, but withdrew when findings from Framingham continued to cite cigarettes as harmful agents. "We turned a blind eye to where Dawber was getting outside money," says Feinleib. The insurance industry was especially interested in keeping the Study alive, because Framingham findings helped it understand the risk profiles of policyholders.

Dawber learned a basic trick of the trade. Tell donors what they want to hear. "In going to these outfits, the first question they asked was, 'Do you have access to all the data?' I had to assure them that I had, although I didn't," he says. That's because the data were officially the property of the federal government, and the National Heart Institute would not give him full access to the data he was instrumental in collecting.

But Dawber and McNamara—a meticulous record keeper—gathered copies of just about everything they could get their hands on. "When we put everything into the computers, we had millions and millions of IBM cards. I saved them all. I thought if anything ever happens to any of this, you can never go back. I saved them in file cabinets downstairs in the cellar," she says. Dawber took the IBM punch cards and, at a personal cost of $10,000, copied all the data to bring with him to Boston University. They both tell the story in a way that is confessional, a secret kept between them well into their eighties, each with the notion of protecting the other, and each agreeing that, at long last, it needn't be kept secret any longer. Neither one even knows if what they did violated any laws. After all, if Dawber had kept his own copies of everything from the start, no one would have wrested them away. But now he had copies of all the data, and he spent the next four years raising $500,000 to keep the exams, and the Study, going.

On one fund-raising trip to St. Louis, hoping to tap some insurance companies for support, Dawber ran into Paul Dudley White. By then, White was known well beyond the medical community. Dawber recalls White's shock at learning that Dawber was traveling the country—essentially with hat in hand—in an effort to save the Study. "He immediately said, 'Put me down for $100.'" Perhaps most touching, about one thousand private citizens of Framingham, hearing that their Study might fold, started sending in checks for $10, $25, $100, and more—donations that added up to more than $40,000. "Roy Dawber was doing groundbreaking stuff. I don't think people realize how great his contribution was. His effort saved us," says Kannel.

By the time the decision had been made to close the Study, two things were in place that ultimately saved it. The core staff positions in Framingham were funded, though they could not continue examinations. Had the Study lost Kannel, McNamara, Castelli, and the connection to Dawber, it would have lost its heart, soul, and institutional memory. It would never have survived. Second was Dawber's connection to Boston University, and the private funding he and the university had set up, thanks to all the data he brought with him. This funding enabled the Heart Study to continue performing examinations.

All this was happening behind the scenes. What the public heard was that the Framingham Heart Study was shutting down. Years of internal turmoil turned into a sweeping public outcry. "I warned Dr. Cooper [then NHI director] that there would be repercussions because of the widely accepted information that was being given to practicing physicians by Kannel's and Castelli's presentations to medical societies around the country," said Zukel. Doctors throughout the country supported the Study because it was giving them information that could help their patients avoid heart disease.

But no one in Washington anticipated the reaction of the general public. Members of Congress gave speeches. Newspapers ran stories. A *New York Times* headline on October 3, 1969, read, "Big Study Program on Heart Attacks to Close Because of Fund Shortage," and the article said that "the plan to close the program was not announced formally, but was noted briefly today in a newsletter called American Patient, published by the American Patients Association, a consumer group. Officials at the Heart Institute confirmed the report."

Letters poured in to newspapers, to the NIH, to members of Congress, to President Nixon. Kannel got into hot water over the public uproar. "The trouble was, the NIH thought we were soliciting all this attention. They were really upset. I personally was called down to Washington, where I sat at a table with all the NIH directors who said, 'We don't like this.' I said, 'Hey, don't look at me.'" They showed Kannel letters of support for the Study they had received and blamed him for playing to public sympathy. The Study's impending shutdown was being protested everywhere. "The protest was so formidable, from so many sources, that the folks at the NIH concluded that we must have been soliciting it. Not at all. The only protesting we did was to the people at NIH when they asked our opinion—which they seldom did!" says Kannel.

"These letters would come in, and they'd bounce from Nixon to the NIH directors and then to Bill Zukel, the project officer. He would send them to me and tell me to explain to the writer why they're closing it. I said that I wasn't going to do it because I didn't agree with the decision. So there I was in Washington, confronted with the whole top-echelon hierarchy explaining to me the position of the NIH, presumably to intimidate me into supporting their position." When Kannel was instructed to write letters saying that he supported the Study's termination, he refused, telling them, in two words he wouldn't want to see in print, exactly what he thought of their idea.

"Technically, I guess I disobeyed orders," says Kannel. He claimed he didn't do anything to feed the furor. "They didn't believe me. They said, 'You're in the Public Health Service, and this is insubordination.' So I've got in my record 'Insubordination' for refusing to justify closing this place down. . . . All sorts of people got involved. There was a huge outcry. But we had contacted nobody. *Nobody.* The hue and cry was based on the merits."

To this day, it's not clear to me whether Kannel was actively involved in gathering support for the Study to stave off its premature closure. At the very least, he was kept well informed by Dawber. Kannel and Dawber admired each other and confided in each other. For two decades they had created and nurtured the Study. Then Washington threatened to close it down. They clearly felt betrayed, and their loyalty rested with the Study and its promise for ever more discoveries, not with bureaucrats in Bethesda.

When Senator Ted Kennedy of Massachusetts heard of the impending shutdown, he contacted Cooper, expressed concern, and asked why the decision was made. Secretary of Health, Education, and Welfare Robert Finch asked the National Heart Institute for a background report on Framingham to use in dealing with high-level inquiries.

Reaction to the decision took on a life of its own, with distinct political overtones. Senator William Proxmire had made a name for himself with "Golden Fleece Awards," granted to government expenditures he considered exceptionally wasteful, but Framingham didn't fall into that category as he read this into the *Congressional Record:*

Mr. President, today's *New York Times* carried a story about the closing down of a major study program of the NHI. The program has been in existence for the past 20 years, and has developed significant new data on high blood pressure, cigarette smoking, obesity, and cholesterol as risk factors which contribute to heart disease.

Now, because some $400,000 and about 20 staff positions have been cut from the Institute's budget, the program will have to shut down, with its work only partially completed.

Mr. President, this makes absolutely no sense to me. How can this country in one breath announce that it lacks the resources to find ways to save hundreds of thousands of lives—heart disease is the nation's number one killer—and in another breath signal the go-ahead for a supersonic transport plane? We cannot find $400,000 for research on heart disease. But somehow we have no trouble raising $1,200,000,000—3,000 times as much—for the SST.

Mr. President, what has happened to our sense of priorities?[13]

One letter of support carried more weight than any other as it made its way from Massachusetts General Hospital to the White House. Paul Dudley White once again spoke up for the Study. He wrote a letter to President Nixon emphasizing the tremendous value Framingham results had for practicing physicians as a blueprint for improving their care of patients. His letter of September 2, 1969, was a reminder that the Study directly affected the lives of every doctor, every patient, and every person in America.

Dear Mr. President:

It has been called to my attention that one of the major research programs on heart disease—the Framingham Heart Study of the United States Public Health Service—is threatened with extinction on July 30, 1970 because of budgetary limitations.

I was a member of the committee which created the Framingham Heart Study Program in the late 1940s and have been closely affiliated with it ever since. A random population has been followed for almost twenty years for all the known possible factors underlying the occurrence of diseases of the heart and blood vessels. For example, data collected by the study have revealed exact quantitative interrelationships between acute heart attacks due to coronary disease, our current very serious epidemic, and such factors as diet, high blood pressure, cigarette smoking, blood cholesterol and overweight. The research program continues to supply basic knowledge on the natural history of coronary heart disease that is available from no other source.

If the study is to be terminated next July, the population will be lost and cannot be retrieved. It will be necessary to start once again and follow another population for several decades before the information now almost available could be collected. The situation is all the more serious because the Framingham Heart Study population is now entering the age group where stroke also is common. If the program were permitted to continue, the factors responsible for stroke in addition to those concerned with sudden death from heart disease could be defined.

The Epidemiology and Biometry Advisory Committee to the National Heart Institute stated, "The Committee strongly feels that the Framingham study should be regarded as a national resource for the study of the natural history and epidemiology of a broad spectrum of diseases which are highly prevalent in the older members of our population" and made a unanimous recommendation among others that "the study of stroke alone would be of sufficient importance to justify continuation of the follow-up of the Framingham cohort for another ten years." They also recommended "a mortality follow-up of offsprings and brothers and sisters of the cohort subjects"—information that is nowhere else available.

The budget of the Framingham Heart Study Program is about 300,000 dollars per year. In the light of the enormous sums that are being spent in other programs such as space, it would seem that the United States Government should without question continue a program so essential to the health of Americans.

I therefore, respectfully request, Mr. President, that you make a personal effort to save the Framingham Heart Study Program.

With all best wishes, I remain

Sincerely yours,

Paul Dudley White, M.D.

The next month, on October 22, 1969, President Nixon replied:

Dear Dr. White:

It was very thoughtful of you to write to me about the Framingham Heart Study. The wisdom of this important undertaking has been confirmed by the immensely valuable knowledge which has come from the program and it is a tribute to your judgment and that of your colleagues in the field of heart disease that the study was initiated some twenty years ago. I appreciate your concern that this research program not be interrupted and I understand that although the Heart Institute will cease to have sole and direct responsibility for the Framingham study as of next June 30, it will continue to make extensive use of this resource, following the Framingham population for documentation of heart attacks and causes of death for as long as significant information can be obtained. You may know that the Heart Institute also plans to assign five staff members for an additional year to complete the task of preparing all the collected cardiovascular data for statistical analysis. I have sent a copy of your letter to Director Mayo at the Bureau of the Budget and I am sure that his office will be in touch with you further about this in the near future.

With warm personal regards,

Sincerely,

Richard Nixon[14]

White replied with a note of thanks, saying that if the Study were to continue, "it would be worth its weight in gold."

"I think it was probably Dr. White's letter to President Nixon that may have at least modified the decision on termination," Zukel says. Nixon was vice-president when White cared for President Eisenhower following his heart attack and was well aware of White's reputation.

A mechanism for compromise was devised. As planned, twenty-two Heart Study staffers lost their jobs, but a core five remained. Dawber had everything in place at Boston University to continue the examinations. Public support and private funds were a bridge that kept the Study exams going. But after four years of fund-raising, Dawber was worn out and felt he couldn't keep up the effort.

Within the National Heart Institute, Feinleib was busy on two fronts. He drafted a proposal outlining a collaborative effort to add to the Study's skeletal federal staff. The proposal resulted in a long-standing partnership with Boston University that began in 1971. He also started to believe that genetic and family factors had been neglected in epidemiology, and worked with Framingham researchers on a plan for a study of the sons and daughters of the original volunteers. He and others envisioned a multigenerational extension as a step to reinvent the Study and direct it down a bold new path. Enrolling Framingham's next generation would yield a wealth of genetic information. "We snuck it in on the side. In preparing the 1970 budget, it went in without paperwork. It was just a footnote," Feinleib recalls. The footnote got funding, and the clinic in Framingham, with its staff and support in the form of an NIH contract with Boston University, could continue. A new generation of volunteers joined what became known as the Framingham Offspring Study.

By 1971, the dispute was over. The Study became a collaboration of government and academia, and remains so to this day. Since 1971, Boston University has been the recipient of a contract with the National Heart, Lung, and Blood Institute to provide examinations and extensive analytical and scientific support.

Had the Framingham Study died in 1968, nearly a thousand projects that followed would not have been possible. For example, a lot of information about women would have remained untapped, because important discoveries such as the role of menopause in cardiovascular disease were not made until the 1970s and 1980s. Like women everywhere, many women in Framingham had begun taking hormone replacement therapy after menopause in the hope of keeping cardiovascular disease at bay. The Study would show that, in fact, it increased their risk for stroke.[15] That finding so flagrantly flew in the face of conventional scientific thinking that it was dismissed for years, until proved beyond a doubt in 2002 by the Women's Health Initiative. When Framingham investigators "found that there was no benefit from HRT [hormone replacement therapy with estrogen] and, in fact, an increased risk of stroke, everybody said, 'Baloney,'" according to Kannel. "It turned out we were right."

The Women's Health Initiative called a halt to a massive study comparing estrogen (plus progestin) with placebo in women following menopause.[16] Researchers discovered that hormone replacement therapy was doing more harm than good. The multicenter clinical trial confirmed the Heart Study's earlier observations.

In the decades since the crisis, the Study has published hundreds of important findings. The participants showed that physical activity reduced the likelihood of heart disease,[17] and that obesity increased it.[18] Atrial fibrillation, a condition in which the heart beats irregularly because of chaotic electrical activity in the upper chamber, increased the risk of stroke fivefold in Framingham participants.[19] Until then, the condition was considered innocuous. Clinical trials resulting from that information tested anticoagulant drugs like warfarin and showed that such medications could prevent strokes in patients with atrial fibrillation.

High levels of high-density lipoprotein or HDL cholesterol actually protected against heart disease in the participants.[20] Although we do not yet have conclusive evidence that raising HDL will reduce a patient's risk for heart disease, we do have suggestive data to that effect. New drugs in development directly raise HDL levels; in the future we will have studies to determine if such drugs are beneficial.

In 1998, Framingham researchers published a risk prediction formula that calculates, using numbers any physician can provide from a standard checkup, a patient's chances of getting heart disease within the next decade.[21]

The unfolding cholesterol saga, like the high blood pressure story before it, is a tale that played out over a period of decades. And unlike getting older, or being male, blood cholesterol levels would become—thanks to basic science, animal studies, pharmaceutical company research and development, and human clinical trials—a characteristic people could at last do something about. The Framingham volunteers were proving that "average" blood pressure and cholesterol numbers, once a source of comfort and complacency, could be fatal. It was a revelation that to have average values for risk factors should not be confused with being in good shape. The average American is no model of well-being.

"What was once considered innocuous blood pressure we showed as a continuous gradient of risk," says Kannel. "Most heart attacks come at moderate blood pressure levels, say 140/90. We kept moving the concept of 'normal' away from 'usual' and toward 'optimal.' " People could be exactly like the vast majority of their neighbors and still develop heart disease. Framingham researchers would emphasize this point again with cholesterol and HDL levels, with obesity, with blood sugar levels, and with sedentary lifestyle.

The Study findings continue to grow more and more specific. The effect of total cholesterol was known, but not the associations of triglycerides and LDL and HDL cholesterol with heart disease. The role of homocysteine in heart disease was more than twenty years away from being studied by Framingham researchers. Lipoprotein little a (Lp[a]) would not be found as a risk factor for more than a quarter of a century.

Few of the naysayers are still alive, and none are in positions of policy-making power. Donald Fredrickson, who served on the Framingham review committee in the mid-sixties and later became director of the National Heart Institute, came clean to Kannel more than two decades after the crisis. "He admitted to me that he thought it was the worst mistake they ever made," Kannel says. What is now called the National Heart, Lung, and Blood Institute (NHLBI, formerly the NHI) showed its appreciation for the Framingham Heart Study at the fiftieth-anniversary celebration in 1998. The institute listed forty-one research milestones from the Study, and acknowledged it as a jewel in the crown of the National Institutes of Health. Of the milestones, all but eight came after 1968. Claude Lenfant, director of the NHLBI, wrote supportively:

> The NHLBI is extremely proud to have shared so much of its history with the Framingham Heart Study. The concept of risk factors, first articulated by the study, has become an integral part of modern medical thinking about chronic diseases. Moreover, the major findings from the study have enabled the [National Heart, Lung, and Blood] Institute to develop and implement effective national prevention and education strategies for clinicians, control of risk factors such as hypertension, high blood cholesterol, and smoking have improved remarkably, and death rates from coronary heart disease and stroke have declined precipitously. There can be no better proof of the value of Framingham's gift to the nation.[22]

And Dawber, keeping it all in perspective at the celebration, referred to the averted shutdown with decided understatement, being every bit as discreet as he had been when writing his book about the Study. He referred to the near death of his beloved Framingham Heart Study as "that one difficulty."

NINE

Coming of Age in a
High-Technology World

The Framingham Study survived and took to heart the criticism of the review committee headed by Abraham Lilienfeld. Its report in 1968 suggested that the Study could not simply continue doing the same exams over and over again. We would need novel measures generating additional kinds of information. The introduction of a second generation of participants in 1971 was part of the answer. It would give researchers the opportunity to explore family patterns of cardiovascular disease. But that wasn't enough. Technology was opening up whole new fields of research. There would be little room for innovation without beginning to use these tools to delve deeper. In Framingham in the late 1970s, we could diagnose coronary heart disease only after a participant developed angina or a myocardial infarction. But symptoms of coronary disease occur late in the progression of atherosclerosis. Technology could help us find signs of disease sooner. That was a central aim of the U.S Public Health Service when it initiated the Study in 1948, but it was destined to fail because screening technologies didn't exist.

The plan to enroll the sons and daughters of the original volunteers became known as the Framingham Offspring Study. Enrolling the 5,124 second-generation participants was nothing like the effort to get their parents involved. This generation was practically waiting for the phone to ring. "As a child, I remember the folks talking about their appointments coming up. Mother was active in the PTA and synagogue, Father was active in town. It was all part of the fabric of this community," says Larry Shapiro, a participant in the second-generation Study. By 1971, when the first of the grown children began coming in for their exams, the Study had become a family tradition.

127

They knew more than their parents did about blood pressure, cholesterol, smoking, exercise, and diet. Most of them are baby boomers, and as they grew up, they could choose to reap the rewards of the information provided by the examinations of their parents. In some ways, they did better. They have lower blood pressure and cholesterol levels and are far less likely to smoke than their mothers and fathers at the same age. But in one major way, they are a lot worse. They are fatter, at younger ages, than the first group of participants.

The core tests the offspring participants underwent in the clinic remained much the same as they were in 1948 when their parents joined the Study. Blood was drawn for laboratory tests. Measures of height, weight, and blood pressure were taken. Skin folds were measured. An electrocardiogram was administered. But their second exam in 1979–83 included much more. Castelli, who directed the Study from 1979 to 1994, hired its first staff cardiologist, Daniel Savage, as well as a statistician with a keen interest in computers, Daniel McGee. They brought with them a different kind of curiosity about studying the heart, and an infusion of enthusiasm for new technology. They were a good fit with Castelli and his newfound, and long-enduring, fascination with computers. He put the Study's first personal computer on his desk, a machine he used mainly for word processing. He lent it to Savage so often that he finally gave it to him. Together, they ushered the Study into the modern age of research.

During my first experience with Framingham as a medical resident in 1982, Savage was my mentor and the person most responsible for igniting my interest in a research career. I would literally follow in his footsteps, taking over his position only a few months after he left in 1983. He had a singular determination to introduce new diagnostic methods to the Study. In his short tenure there, he brought in treadmill exercise testing and ambulatory electrocardiographic monitoring, making for a more active exam. At first, participants would be wired up to the ambulatory device for an hour or two during their exams. They were encouraged to wear them overnight, returning the equipment and tapes the next day.

Savage also introduced echocardiography, enabling the Study's researchers to identify heart disease in its early stages. He and his wife, Sandra, the echocardiography technician, took on the new studies with vigor. "They [participants] were coming in at the rate of four an hour," Sandra Savage remembers. "In five years, we did eight thousand of them." Subjects lie on their sides with a transducer, a sound wave transmitting and receiving device shaped like a microphone, held at their chests while ultrasound waves create moving pictures on a monitor screen. Some like to look up at the screen and watch their hearts beat. Dan Savage kept pushing to do more. "At first it was, 'Let's try it out,' to 'Let's do some more,' to 'Let's do everyone,' " she says.[1] By the time the sixteenth round of exams for the original cohort began, thirty-two years after the outset of the Study, all the participants were getting echocardiograms.

Dan Savage was a go-getter with boundless energy and a drive to get answers. In Framingham, he was frustrated by the decidedly slow pace of epidemiology. "Here, you take the measurement and you wait five years to see how it plays out," says Castelli. It was a pace alien to Savage, who worked long hours, pulled regular overnighters, and carried three or four briefcases with papers and data relating to different projects. He was constantly planning new tests and figuring out what new equipment he'd need to carry them out.

Technology had altered the Study irreversibly. Castelli, propelled by the unbridled enthusiasm, ambition, schemes, and machinations of Dan Savage and, to a lesser extent, Dan McGee, was putting in so many requests for equipment through the slow federal process within the Heart Institute that Robert Levy, then director, quietly gave him the okay to set up a surreptitious account to bypass the usual formalities and waiting periods for equipment purchases. "I said [to the director] I was tired of asking permission every five minutes," recalls Castelli. "I asked him what the limit was to what I could order without permission." There were knowing glances, but there was no answer to the question. So they set up a fund, referred to it as the "paper clip account," and spent amounts that might have supplied half the country with paper clips. "We bought all this fancy equipment through the limitless paper clip account," Castelli says.[2]

Savage helped usher in the technological era, but he left for the Centers for Disease Control long before his ideas could be fully tested. He died prematurely just as the data from the tests he introduced in Framingham were bearing fruit. He also never lived to see the seeds of

er idea germinate. In 1995, we began recruiting a minority popu-
for what we call the Omni Study, to understand why heart disease
disproportionately in some groups and to find out if risk factors
are equally common and dangerous among minorities. Although the
Omni project began a decade after Dan Savage left the Framingham
Study, as a member of the Association of Black Cardiologists, his interest in such a project ran deep. He started recruiting minorities in Framingham in the early 1980s, before such an obvious idea got the approval
of the NHLBI. He kept a carton of files with charts on about one
hundred minority volunteers from the community, and those people
became the first members of the Omni Study.

The new diagnostic tests that excited Savage changed the very
nature of the Study. Framingham research, for example, entered into
the debate over the meaning of a positive treadmill test in asymptomatic
people. Many believed that there were a lot of false positives in such
results. "We picked up people with a positive treadmill test who otherwise looked perfectly normal," according to Castelli. Eventually, we
showed that an abnormal treadmill test is associated with increased risk
for developing angina or a heart attack, although false positives do
occur.[3] We found that how long someone could endure on the treadmill was a better predictor of heart disease than the exercise electrocardiogram. Other than performing an invasive angiogram, which involves
threading a catheter up the aorta into the heart to look for damage to
the coronary arteries, the exercise test was the first widely available tool
for diagnosing occult coronary artery disease.

And the ambulatory electrocardiographic monitoring that Savage
introduced allowed us to analyze cardiovascular health as people went
about their normal daily routines. Those results helped us conclude a
decade later that in people without apparent heart disease, the occurrence of ventricular arrhythmias is associated with twice the risk of heart
attack.[4]

The most important and enduring discoveries from the newly introduced tests came from the echocardiograms over which Dan and Sandra
Savage toiled. When the Study began more than fifty years ago, we
were able to diagnose heart problems, such as myocardial infarction,
only by their symptoms or telltale patterns on the electrocardiogram.
But by then the disease is far advanced. Echocardiography provided a

direct and early anatomical diagnosis of heart disease in asymptomatic people. We could look at and describe what healthy heart chambers look like, what normal heart wall thickness is, what the total muscle mass of the left ventricle should be. Using the echocardiograms that Dan and Sandra Savage performed and measured, the Study found that left ventricular hypertrophy, a thickening of the muscle wall, was far more common than anyone thought.[5] We discovered that even small elevations in blood pressure over a period of years were associated with thickening of the walls of the heart.[6] Those two pieces of information suggested a need for early and aggressive hypertension treatment.

With several years of follow-up, we learned that people with evidence of thickening of their heart muscle were at substantially increased risk for the development of heart attacks and sudden death.[7] The Study reported that enlargement of the left ventricle is a risk factor for congestive heart failure, improving the ability to identify people at risk for that dreaded disease.[8] Echocardiography showed for the first time that changes in the structure or function of the heart could predict the risk for atrial fibrillation, a potentially dangerous heart-rhythm disturbance.[9] It also proved to be an added tool for predicting risk of stroke.[10] The echocardiogram gave us a window on the heart and allowed us to look with great precision at the damage caused by hypertension. Clinical research brought the good news that controlling blood pressure can reverse the damage to heart muscle, and that early treatment of a dilated heart can prevent the progression to heart failure.

The changes introduced by Castelli and implemented by Savage in the late 1970s weren't limited to the clinic exams. The way data were managed was also altered forever. For about thirty years, the Study had been like a well-oiled machine, running predictably year after year. And all those years, Pat McNamara had been the overseer of virtually every number recorded in mountains of data. Her job was to guard the contents of the Study, making sure that everything was done precisely and that nothing changed. Data-related procedures, and how to implement them, had been housed in her institutional memory. By the early 1980s, the changes going on around her were threatening and overwhelming her precise and organized mind. Change was barreling through every aspect of the Study, and computers would prove to be just as fastidious as—and far more rapid than—Patricia McNamara.

When Dan McGee came on board, so did the introduction of on-site computers in a form that already seems primitive two decades later. The change was launched with a series of memos from McGee in 1983.[11] Today, it seems just as incredible that a project that is all about data existed for thirty-five years without computers as it does that Franklin Roosevelt's escalating blood pressure went untreated just a few years before the Study began. McGee's plan was to have stand-alone computer workstations that recorded data, with information collected on floppy disks to be uploaded later. Eventually, data from the eighteenth cycle of examinations for the Heart Study participants were entered into a computer system in Framingham. No longer would we mail forms to Bethesda for keypunching and receive the data back. Today, of course, we have sophisticated on-site data entry and computer programs with built-in checks to pick up any suspicious information immediately. Starting in 1984, at about the time I joined the Study full time, record keeping was no longer dependent on McNamara's omnipresent pair of eyes. And just as technology threatened secretaries and their shorthand skills or librarians and their card catalog filing systems, it seems to have represented an unwelcome change to McNamara and the work she had always performed.

For decades, she, Dawber, Kannel, and later Castelli were a team charged with maintaining the status quo in order to continue the Study. Now, the tedious manual processing of data, which served the Study well for so long, had become obsolete. Shortly after McGee's 1983 memos circulated, McNamara took several weeks off, something unprecedented in her more than three decades with the Study. End point reviews, or the adjudication of deaths and cardiovascular disease events in the participants, were her responsibility, and these were suspended during her absence.

She returned, but never quite made the adjustment. The abrupt and irreversible change in business as usual was simply too much for her. No longer did she have sovereignty over every number that was entered. Perhaps her discomfort with new technology eased her out of the job she loved, or perhaps she was simply ready to leave. What once took hours of work, poring over IBM keypunch cards, could now be done with a few keystrokes. One day in February 1984, she packed up her desk and prepared to make the fifty-mile commute to her home in

Haverhill, Massachusetts, for the last time. But even facing retirement, she felt a sense of ownership of the Study's information so strongly that part of what she packed was boxes of data. Only years later, with Castelli's intervention, did she relinquish those documents for use in a research project on silent heart attacks.

Although Castelli was McNamara's peer, he was excited by the possibilities presented by new technology. And the introduction of echocardiography, ambulatory electrocardiographic monitoring, and treadmill exercise testing, along with ability to analyze mountains of data in seconds, not hours or days, changed the way research was conducted at Framingham. The dream of some of its founders, to screen large numbers of people for heart disease, was not possible in 1948. There were simply no tools for early diagnosis. With the introduction of new technology, it was at last possible to pinpoint heart disease in its early stages in healthy participants.

TEN

Blood Pressure:
More Than Just a Numbers Game

No one could save Franklin Roosevelt from the internal ravages wrought by escalating hypertension. That scenario continued for millions throughout the 1950s, even as the Framingham Heart Study was gathering the first clues about causes and effects. But it wasn't alone in its quest to understand. Scientists from far and wide were seeking answers during an era when all anyone could do in the face of blood pressure numbers that rose almost inevitably with age was to watch and worry. And because scientific thinking at the time said that rising blood pressure was essential to the aging process, few wanted to interfere with the process for fear of doing even greater harm. Ed Freis, working in Washington, D.C., with funding from the Veterans Administration, was among those most interested in clues from the Heart Study about blood pressure. Clearly, his were among the most daring theories about what needed to be done next.

The Framingham Study was instrumental in proving that hypertension was far from "benign" or "essential," as it was once labeled in medical textbooks. Those beliefs were challenged in 1957 when researchers, including Kannel and Dawber, published results showing that after only four years of observation, Framingham men aged forty-five to sixty-two with elevated blood pressure had an increased risk of coronary heart disease compared with their nonhypertensive neighbors.[1] After six years, the same could be said of women, and of younger men. By 1959, with the publication in the journal *Hypertension* of Kannel and Dawber's paper "Blood Pressure and Its Relation to Coronary Heart Disease in the Framingham Study," elevated blood pressure was proved clearly bad. Far from protecting the elderly by forcing more blood through aging arteries to organs, hypertension put millions at risk for heart attacks and strokes.

Those were among the major findings Freis needed to justify his own farsighted undertaking. Identifying high blood pressure as a serious problem was only the first step in solving the problem. Though he had no formal connection to Framingham, its work gave him solid numbers linking rises in blood pressure with increased risk of heart disease. Freis took the revolutionary step of using the data, available to scientists around the world as they fashioned their own hypotheses, to design an experiment testing whether lowering blood pressure would reduce risk for cardiovascular disease, have no effect on outcome, or possibly do more harm than good. Funded by the Veterans Administration, it was a seventeen-hospital study that first examined the benefits of treating malignant hypertension in 143 men, and then looked at whether reducing blood pressure in 380 men with moderate hypertension would save lives.

It was time to pass the baton from epidemiology research at Framingham to a controlled investigation, or what would come to be called a clinical trial. Today, randomized, controlled double-blind clinical trials carry the ultimate weight of scientific proof. By the early 1950s, Freis was ready to accept the challenge that would come at the end of the decade. He had already begun treating hypertensive patients with chlorothiazide, the most acceptable of available antihypertensive drugs because it was an effective diuretic and did not cause serious side effects. But what he was doing was considered controversial, risky, even unethical, because treating high blood pressure had not undergone the rigors of scientific proof. But now he could move forward. "Framingham supplied me with evidence that blood pressure elevation was a risk, that elevated blood pressure increases the risk of cardiovascular complications," Freis says.[2]

It's one thing to know that high blood pressure is bad. It's another to prove that lowering it makes a difference. Freis was at the forefront, ready to test the theory he had held for more than a decade, that lowering blood pressure saves lives. Yet he remains somewhat conflicted, and a little embarrassed, by the stellar achievement of his life. He proved, for the first time and beyond any shadow of doubt, that lowering blood pressure reduces risk for cardiovascular disease. Today, 50 million Americans have high blood pressure, and a growing number are able to control it, largely through the use of antihypertensive drugs.[3] Freis's research has prolonged and improved the quality of millions of lives. Freis would use the Heart Study's evidence to test the next logical question: Does lowering blood pressure prevent disease?

His internal conflict stemmed from testing a hypothesis he already believed to be true. Was it ethical to withhold drugs to lower blood pressure in order to conduct a clinical trial? When he set out to affirm that lowering blood pressure with the use of drugs would prevent heart attacks and strokes, he believed he already knew the answer. He and a handful of physicians were already treating patients with very elevated levels. His involvement in the Veterans Administration trials of blood pressure treatment is the accomplishment of a lifetime, but he feels a need to defend it.

Because of what Freis established, millions of Americans are able to manage their blood pressure level and, if it is elevated, lower it with drugs to less than 140/90, or even below 130/85. Yet Freis was reluctant to tout his early study results because he was troubled by the belief that the untreated men in the study who died, a calculable body count, would have benefited from treatment. He believed it all along, yet it was only a hypothesis. Men and women across the country were left untreated for hypertension, and neither Freis nor any other scientist could change that standard practice with a hypothesis.

In medicine, a mere belief in a treatment without proof of benefit cannot change the prescribing practices of physicians. Belief is not fact. Intuition, anecdotal evidence, and educated guesses do not add up to scientific proof. "When I came into the picture in 1946, drug treatment of hypertension was considered to be a foolish procedure, verging on charlatanism," Freis wrote.[4] He himself was prescribing medication only to emergency cases. Only the most severe hypertension was treated, and it was often a case of the treatment being worse than the disease. Veratrum vivride, a form of mistletoe that grew in the mountains of North Carolina and was once used by Indians in rites of passage, could bring blood pressure down to abnormally low, even life-threatening, levels. It also caused projectile vomiting. Ganglionic blocking agents, newly available in 1948, could lower blood pressure, but they blocked the whole autonomic nervous system, and side effects included excessive blood pressure lowering, dry mouth, and impotence. Patients saw no reason to accept the drug's consequences, since high blood pressure itself was silent. Sufferers often felt nothing until the moment of disaster.

An antimalarial drug used during World War II, pentaquine, was found to lower blood pressure in high doses—but the treatment wasn't worth the risks. "It was the most toxic drug. It turns you blue," Freis says. "It was too toxic for practical treatment." It caused methemoglobinemia, which made red blood cells take on a bluish hue as it deprived the body of essential oxygen. When patients took it at the dose level required to reduce blood pressure, they experienced severe abdominal cramps and diarrhea. It also damaged the sympathetic nervous system. Mercurial diuretics depleted the body of salt, which lowered blood pressure, but they were toxic to the kidneys. Phenobarbital calmed patients down enough to make it look as if something good were happening, but it didn't lower blood pressure effectively. And the rice diet had simply proved unpalatable. It was a time of early experimentation: the best result was trivial blood pressure lowering, the worst, death.

Before the Framingham research was published, Freis presented his personal experience using chlorothiazide on a handful of patients in the mid-1950s at a meeting of the American Heart Association to a few believers and many skeptics. "I had already had years of experience lowering blood pressure with drugs. Nothing terrible happened. They got better," says Freis. But that was only anecdotal experience. Even today, when more than 25 million Americans with hypertension take medication, its causes remain unknown in 90 to 95 percent of cases. Freis left the meeting feeling certain that drug treatment to lower blood pressure was beneficial to patients. But he also knew that to persuade doctors to change, he would have to prove it.

The arguments over the value of treatment continued for years. The Framingham finding that hypertension led to heart attacks and strokes didn't settle the larger question of whether treatment was harmful or helpful. Standard textbooks advised against treating hypertension. Even as late as 1967, the *Cecil-Loeb Textbook of Medicine* expressed the leading view: "Be sure that the patient really needs treatment. Those over 70 years rarely do, whatever the level of pressure, and certainly should not be treated unless a definite indication such as pulmonary edema, angina pectoris, severe headache or marked shortness of breath on effort is present. . . . Age needs no additional therapeutic hazards."[5]

As the arguments continued, Freis and a handful of clinicians, despite the criticism of peers, treated their hypertensive patients with chlorothiazide.[6] "I had been to a meeting, a panel discussion, where I had to defend the drug treatment against a guy who was very much against drug treatment," Freis remembers. "After the thing was over, I said to myself, I know it's true. But the only way I can really prove this is with multiclinic trials." Scientists opposed to treatment were, Freis says, "big mucky-muck professors" in prestigious institutions. When they spoke, their colleagues listened. Some went so far as to tell Freis he was poisoning people with drugs. It would fall to him, and colleagues of the Veterans Administration Hypertension Study, to turn a controversial belief into a medical fact.

If the chorus of opposition remained even a decade later, one can only imagine the skeptical climate that greeted Freis's 1956 proposal for a cooperative study on the value of antihypertensive drugs.

There were fewer, but equally powerful, voices arguing adamantly against Freis's idea for the exact opposite reason. Those who believed Freis was right about the benefits of treating very high blood pressure were critical of his methodology. Horace Smirk, a New Zealand physician and among the leading advocates of antihypertensive drug treatment, listened to Freis at a 1961 meeting on hypertension in Prague. Smirk's reaction is seared in Freis's memory: "He said, 'You're going to use placebos! Well, you are entirely unethical.' He didn't think it needed proof. He was already treating people."

Freis held tight to the belief that millions with high blood pressure could also be damned if he didn't bring some scientific proof to answer the question of whether or not to treat hypertension.

What the world needed was a definitive clinical trial, the gold standard of evidence-based science. It required an objective comparison of a group treated with antihypertensive drugs and a group of untreated patients.

As chief of the Medical Service at the Washington, D.C., Veterans Administration Hospital, Freis was joined by a dozen or so researchers in starting a controlled investigation. It was an era when the concept of clinical trials was just being invented. "There were no long-term multiclinic controlled trials in the hypertensive literature. We were the first," according to Freis. It was remarkable for its time, using double-blind randomization. That means that the 143 patients were sorted into two groups. One would receive placebo, or sugar pills, and the other would get antihypertensive drugs. Neither patients nor researchers would know which men were in which group. Today, such an investigation, called a double-blind placebo-controlled clinical trial, is considered the strongest research proof. But at that time it was a brand-new way to test a theory.

Freis and his colleagues had little in the way of precedent to design their trial. The first randomized clinical trials had only taken place in 1946.[7] That's when two efforts began, one testing the efficacy of immunization against whooping cough, the other testing streptomycin for treating pulmonary tuberculosis. Both attempted to "blind" the researchers as well as the participants to the treatment type. Until then, medicine advanced when a physician departed from standard treatment and applied a new approach, then compared the success rate with past results. That method could not control for equivalent treatment or for varying economic and social conditions of patients. Nor could it take into account possible differences in severity of the disease. The comparisons were crude and the results were open to doubt.

Well into the 1950s, there were still no institutional review boards to consult, no ethical guidelines written down for the design of clinical trials. Doctors of the era received unqualified respect, and for the most part patients did what they were told. "There was no such thing as informed consent," says Freis. "We told [the patients] they were going to be part of an experiment. In fact, they were eager to participate." It was a far cry from today's rigid standards, safeguards, and layers of oversight and approval to make sure participants in a clinical trial understand fully what they're getting into—a process we now call informed consent. One of the hardest concepts to explain is that the rules of experimentation are made to serve science while safeguarding the interests of the individual patient, and that volunteering comes with no promise of cure or improvement. The rules are well laid out, with layers of consent and review by committees, including physicians and ethicists. The primary intent is to gather information and to move medicine forward for the benefit of humankind, though not necessarily for the individual benefit of study participants. The VA trial began in a different era, when patients asked few questions and trust in physicians and in medicine was close to absolute. Ultimately, the results would benefit countless millions of patients like those in the trial.

Little more than a decade earlier, residents of Framingham had to be persuaded to join an epidemiologic study. But they were perfectly healthy people, not medical patients, asked to have exams though they had no apparent medical needs. In contrast, Freis's recruitment was easier, though the ethics were more complex. He and his colleagues were blazing a new trail and learning as they went. The moral pros and cons of randomized clinical trials were still heavily debated.

Richard Doll wrote in the *British Medical Journal* in 1998 that the streptomycin trial settled the issue of whether it was ethically justifiable to withhold a promising but unproved drug from any patient. Until then, the drug had proved valuable in animal studies and in clinical case reports of treatment of people. "The Medical Research Council's Streptomycin in Tuberculosis Trials Committee agreed that 'it would have been unethical not to have seized the opportunity to design a strictly controlled trial, which could speedily and effectively reveal the value of treatment,'" Doll wrote.[8] So, in a result that showed a remarkable reduction in mortality from TB, the study set an ethical rule when it clinically went up against the old standard of care.

In the case of hypertension, the old standard of care was no treatment at all. Informed consent, in days when doctors were not questioned, didn't even come up. At the time that Freis began discussing his ideas with colleagues, one of the largest and most famous clinical trials ever undertaken was just reporting results. Across the country, 623,972 schoolchildren were injected with a new polio vaccine or with a placebo. The results appeared in 1955: the Salk vaccine was 80 to 90 percent effective in preventing polio.[9] Soon, all children were vaccinated, and the disease receded to near-eradication.

But what began as a simple study to test placebo, or an inactive agent, against chlorothiazide, the best treatment for hypertension known at the time, quickly got complicated. Every physician whom Freis wanted to participate in the study had his own opinion.

"The plan [for a clinical trial] was made out by the doctors. There was no help yet at that stage from any statisticians, and it was a lousy plan," explains Freis. Pretty soon it was untenable. "We were comparing different drugs at the same time we were studying effectiveness and mortality. Everybody was contributing ideas. There was no discipline. What I proposed was a straightforward study—treat some, don't treat the others, and see how they do. What they wanted to do was to try different drugs, compare the effectiveness of different drugs. We piddled around with that for about four years until we got a good biostatistician."

Too many doctors were spoiling the study, and through the late 1950s, the researchers failed to launch a workable placebo control study design. They asked too many questions of too few patients. The first study helped point toward some potentially effective ways to control hypertension, but did not answer the basic question that needed answering: Does control of hypertension prevent cardiovascular disease? That question continued to engender ethical attacks from all sides. The vast majority of the medical community thought that treatment of hypertension would do no good, and could, indeed, do more harm than good. Big names in cardiology, like William Goldring and Herbert Chasis, believed it was unethical to treat hypertension until medical science understood the underlying cause, a search they believed should be undertaken in the laboratory. They wrote with certainty: "There can be no doubt that the disabling and lethal agent in chronic hypertensive disease is not the blood pressure but the associated arterial and arteriolar disease."[10]

Reducing blood pressure, they argued forcefully, did no good because it did not affect the fundamental and unknown cause of the disease. Their opinions were medical gospel and dominated clinical practice. So when a dozen or so hypertension researchers met in Atlantic City in 1962, they were prepared to question mainstream medical opinion. On one side of the debate was the credo: Find the cause and the treatment will become obvious. But Freis and his colleagues had a minority opinion: Control the hypertension and you control the disease.

From their earlier efforts, they had learned not to overload the trial with too many questions. They were still on uncertain ethical ground as they hashed out ideas for a simpler, more effective study. They had few resources. Grants were going to laboratory scientists looking for causes of hypertension. Freis tells of the financial pinch:

We were told we could go ahead and do a study but they didn't give us any money. So I got in touch to see who was going to be there and suggested we get together at lunchtime at the Seaview Hotel because it had a second-floor lobby that was always empty, and we didn't have any money to reserve a room. You know the trials that are carried out today cost tens of millions of dollars. Ours cost something like $50,000 a year. How did that happen? We all contributed our time. We didn't take any money at all. We were all VA employees, and we didn't get any extra money for contributing.

Finally, in 1963 Freis and a core group of investigators from the earlier, aborted study planned a trial designed to answer just one question: Does treatment of moderate-to-severe hypertension reduce death, stroke, and heart disease? They needed a statistician, and got one in Lawrence Shaw, a pioneer in the new field of biostatistics. Shaw wasn't intimidated by doctors. He set the scientific agenda and steadfastly kept the focus on the single mission. His mantra was to keep it simple, and he designed a trial that to this day remains uncriticized for its methods. "It was a straight design that you couldn't mess up," says Freis. "Then Shaw maintained discipline among us. He didn't let us break protocol. If urine samples weren't sent in, he would let you know about it. And he kept a close tab on the results." Today, such procedures are standard, but back then they were groundbreaking. They tested placebo against what they hoped to be the best regimen of the day, a combination of chlorothiazide and two newer antihypertensive drugs on the scene, reserpine and hydralazine.

They chose veterans whose diastolic pressure, the lower number, was between 90 and 129 after a few days of hospital rest and monitoring. That spanned a spectrum from mild to severe hypertension. The group included 143 men with severe hypertension: diastolic pressures between 115 and 129. Their average age was fifty-one, and they weighed, on average, 184 pounds. Seventy-seven were black and sixty-six white. The diastolic pressure in the treated group fell from an average of 121 to 91.6 after a year of treatment. That of the placebo-treated group remained at an average of 121.[11]

Shaw, who was the only one privy to the results in the "blinded" study, sounded the alarm in 1967. An early analysis of results showed that patients with severe hypertension on placebo were faring much worse than those receiving active treatment. Serious cardiovascular events had occurred in twenty-seven of the placebo-treated patients compared with only two on active treatment. The study of men with severe hypertension was halted, and patients on placebo were prescribed hypertensive treatment.

Four patients in the placebo-treated group died, three from ruptured aortic aneurysms and one from sudden death. "We regretted that we could not have recognized and terminated this group of patients from the trial sooner, but the events occurred so rapidly they caught us by surprise," Freis wrote in 1990.[12] Twenty-three others in the placebo group developed complications, including malignant hypertension, congestive heart failure, stroke, heart attack, and kidney failure. In the treated group, one person suffered a nondebilitating stroke. Another dropped out because of an adverse drug reaction. The study began in April 1964 and was to continue until 1969, but never made it that far.

The paper on severe hypertension was published in the *Journal of the American Medical Association* in December 1967. It might have been a time for celebration. "I was thrilled that we were vindicated. But we didn't have money enough to do anything. We couldn't buy any champagne," Freis recalls. Besides, he was reluctant to face those who criticized the group for giving half the patients placebos. "We didn't call anybody. There was no notice to newspapers. I didn't want to. I had a bad feeling that a lot of people thought maybe it was unethical to have a placebo group. I would get criticized because the results for the untreated group were so horrendous." And so, more than thirty-five years after the landmark study was published, three decades after a federal program was formed to teach the public about blood pressure, with millions of Americans choosing from more than sixty medications to control hypertension to prolong their lives, Freis still feels the sting of injustice and misunderstanding at being a pioneer. "Good God, that they would blame me. That I would be criticized for doing the study in the first place, and then be criticized for the results that were positive! I was criticized on both sides."

It still makes some of his colleagues angry. "You cannot criticize someone for not knowing what we know now, like Ed Freis. That really gripes me," says Roy Dawber. And Edward Frohlich, a leading hypertension specialist, recalls an international meeting of cardiologists many years after the VA hypertension results were published. "I remember in a program in London some Englishmen were criticizing the study because it didn't have the numbers, it didn't have this, it didn't have that. I said, 'That's like criticizing Henry Ford for the Model T because it didn't have automatic transmission.' Freis came up with the idea of a multicenter study, in a university setting, double-blinded, with placebo control. No other study had done this."

They had proved that those with severe hypertension benefited from treatment. And about that time, William Kannel was examining statistics from the Framingham participants showing that high blood pressure is also a risk factor for stroke. Drawing on Freis's findings, Kannel predicted that lowering blood pressure would reduce the incidence of strokes. In a 1996 "Landmark Perspective" in the *Journal of the American Medical Association,* Niels Lassen wrote: "Extrapolating this to a far wider group underlies the optimistic tone of the discussion: now we know how to prevent a sizable fraction of strokes. Was this prediction correct? The answer is yes, verified slowly over the years in numerous controlled therapeutic trials."[13]

This is the classic working of science. Kannel's 1959 hypertension findings[14] from the Framingham Heart Study gave Freis the information he needed to design a clinical trial. But the favor was returned. Kannel used Freis's initial results to go out on a limb and predict that strokes likewise could be prevented by blood pressure control.

But what about less severe hypertension? The VA Cooperative trial monitored differences between active drug treatment and placebo for the men with less severe diastolic pressures of 90 to 114,[15] until October 1969, when Thomas Chalmers, an authority on clinical trials, looked at the latest statistical analysis. That arm of the study was also stopped prematurely because of clear-cut benefits of treatment, and all patients were put on active therapy. Freis had proved that blood-pressure-lowering medication also helped mild to moderate hypertension. Treatment reduced the risk of a cardiovascular event. In human terms, nineteen men in the placebo-treated group died of causes related to hypertension, while in the actively treated group, eight men died. The five-year, seventeen-hospital study established once and for all that drug treatment for moderately elevated blood pressure is effective in preventing major complications like strokes, congestive heart failure, and kidney failure. It was only after publication of the second VA study that the dominant medical opinion was finally proved wrong.

"When it started coming out that treatment even helped people with moderate hypertension, I started thinking, 'This really involves the health and welfare of millions,'" says Freis. "By then, I wasn't worried at all about criticism. What was going through my head was, 'Here's something we should publicize right away.' I called a press conference. We had people from the various news services, the AP, UPI, the *New York Times.*"

The syndicated AP report made the inside pages of a few newspapers. Walter Cronkite mentioned it on the CBS *Evening News*. Freis began to get invited to speak to groups of physicians around the country. Soon, he was swamped with speaking engagements. His primary work switched from research to teaching physicians. He spoke and answered questions about the VA study, the treatment they used, and the results they got. He wanted to change the behavior of doctors so they could change the lives of their patients.

In 1971 Freis spoke at a seminar on clinical trials held by the American Society for Clinical Investigation, and Mary Lasker heard about his talk. She was a philanthropist who used her social and political connections to encourage medical research and raise public awareness about diseases. She and her husband, Albert, created the Lasker Foundation, honoring outstanding discoveries in basic and clinical research. She would visit governors, legislators, congressional leaders, and presidents to push and prod for more funding for medical research. She used to say, "I am opposed to heart attacks and strokes the way I am opposed to sin."[16] She went to bat for Freis and his research on hypertension. "Dear old Mary Lasker. She went to Elliot Richardson [then secretary of Health, Education, and Welfare] with reprints of my papers and publications. She said, 'Now look, here it is and it has been proven, the importance of the treatment of hypertension and the doctors are not applying it,'" says Freis. She caught Richardson's attention, in part because the secretary's father had died of a stroke due to hypertension. And that led to the 1972 establishment of the National High Blood Pressure Education Program, designed to measure and treat high blood pressure.

Other studies over the years would continue to confirm that lowering diastolic blood pressure, even when only mildly elevated, was beneficial. Later research would prove that women, like men, benefited from antihypertensive treatment. Even the elderly, who were once thought to owe their very survival to "essential" hypertension, lived longer and with fewer heart attacks and strokes if their condition was treated. William Kannel wrote in 1970, "Asymptomatic, mild hypertension is far from innocuous even in the elderly."[17]

Whereas in young hypertensive patients the diastolic blood pressure is elevated, in older people the problem is a high systolic pressure. While Freis's study looked at the benefits of treating diastolic hypertension, much less was known about treating systolic hypertension.

Even as recently as 1991, a lot of doctors didn't believe in treating elderly patients with high systolic blood pressure. The Systolic Hypertension in the Elderly Program (SHEP) showed that year that treatment of the ailment in patients sixty years of age and older is every bit as effective in preventing cardiovascular complications as it is in younger people with diastolic hypertension.[18] Finally, after years of concentrating on diastolic blood pressure, results of the SHEP study confirmed findings from the Framingham Heart Study showing that systolic blood pressure was equally important. It was now certain: Elevated systolic pressure should also be treated. Howard Bruenn, FDR's concerned and frustrated cardiologist, could have worked wonders with today's information and drugs. Yet despite all the evidence that treating older patients with systolic hypertension is beneficial, even today only a small fraction of them have their blood pressure treated and controlled.

The germ of an idea developed into a conviction for Freis. The conviction led him to a battle of intellect with the medical establishment, and the fight was won by scientific proof. It took twenty years.

Chlorothiazide was the first answer, and diuretics are still used as a first line of defense against hypertension. But their use has dropped considerably since the advent of newer and more costly drugs like calcium channel blockers, ACE (angiotensin converting enzyme) inhibitors, and angiotensin receptor blockers.

Some thirty-three years after Freis found that chlorothiazide controlled hypertension, a similar diuretic, chlorthalidone, now old enough to be available as an inexpensive generic drug, went head-to-head with three newer medications in the largest hypertension clinical trial ever conducted, with 42,418 participants including men and women, Hispanics and blacks. Each class of newer drug used in the trial had been approved after studies showed it lowered blood pressure better than placebo. But they had never been tested head-to-head against diuretics.

In that comparison trial, known as the Antihypertensive and Lipid-Lowering Treatment to Prevent Heart Attack Trial, or ALLHAT, supported by the National Heart, Lung, and Blood Institute, the newer drugs were found to be slightly less effective in controlling blood pressure and considerably less effective in preventing heart failure than the diuretic chlorthalidone.[19] And Claude Lenfant, NHLBI director, said after completion of the study, "these more costly medications were often promoted as having advantages over older drugs, which contributed to the rapid escalation of their use."[20] In fact, diuretic use fell from 56 percent of antihypertensive prescriptions in 1982 to 27 percent in 1992.[21]

When results were published in the December 18, 2002, issue of the *Journal of the American Medical Association,* they were touted in a way that stands in stark contrast to Freis's humble, almost embarrassed, response to his study results. Following ALLHAT, national guidelines were changed, a press conference was held, and major newspapers, including the *New York Times* and *Washington Post,* carried front-page stories with detailed reports on the findings.

"ALLHAT shows that diuretics are the best choice to treat hypertension and reduce the risk of its complications, both medically and economically," Lenfant said of the findings.[22] A regional coordinator of the study, Paul Whelton, senior vice president for health sciences at Tulane University, added, "ALLHAT's findings also indicate that most patients will need more than one drug to adequately control their blood pressure, and one of the drugs used should be a diuretic."[23]

A year before the study's results were published, Freis told me he was still frustrated with his colleagues and the way they were treating high blood pressure. "I think the field has gotten confused. I am not happy with the current crop of hypertensive experts. They are pushing monotherapy and that you design therapy according to the co-morbid condition. They get poor control. I think monotherapy is not as effective as two-drug therapy. We controlled blood pressure almost 100 percent in the VA study by combining a diuretic with other drugs. We could control blood pressure much better than we are doing currently." Once again, his thinking has been vindicated.

Today, there are dozens of drugs available to treat high blood pressure, and eleven new ones are being developed, according to the Pharmaceutical Research and Manufacturers of America. But, as the saying goes, what goes around, comes around. Despite the many new drugs available, and $15.5 billion spent annually on medications to lower high blood pressure,[24] the ALLHAT study proved that diuretics, the drug class that was the cornerstone of Freis's clinical trial nearly four decades ago, were superior in blood pressure control and in preventing some of the most serious adverse consequences of cardiovascular disease than newer and much more costly medications. Using thiazide diuretics in the treatment of the 50 million people with hypertension in the United States would save billions of dollars each year. That money could be better spent on further research into the causes of heart disease and stroke, or applied to additional clinical steps to prevent heart disease in those at greatest risk.

The tools exist to help, yet 30 percent of those with hypertension today are unaware of their condition. Overall, only 34 percent of adults with hypertension have their blood pressure controlled to below 140/90.[25] Most of them feel no symptoms, no obvious daily reminder that something is wrong. Those who come to their physicians' attention usually begin trying to control their hypertension with a weight-loss diet, salt reduction, and exercise. Unfortunately, Americans are notorious for failing to maintain a consistent diet-and-exercise regimen.

But even when they get a prescription for an antihypertensive drug, many patients simply don't take the pills consistently. Noncompliance is a major reason so many Americans continue to have high blood pressure. How do you motivate individuals, in apparent good health with no symptoms, to believe that taking a pill every day, indefinitely, is prudent? So many feel they'd rather do it "naturally," apparently not realizing that the "natural" history of uncontrolled hypertension is disastrous, while today's drugs are safe and effective. If side effects develop, as occasionally occurs, there are many other drug options available.

It is possible to improve upon our current low rates of blood pressure control in the United States. Once we identify those with hypertension, we must also be more thorough in starting treatments—both lifestyle changes and drugs—for those with persistent blood pressure elevation. But the task is more complex. Additionally, we must develop methods to keep patients on beneficial treatment after it is started. A study reported in the *American Journal of Public Health* followed 400 hypertension patients. Half of them were counseled about the consequences of high blood pressure, and a family member was recruited as a kind of benevolent nag, reminding them to take their medication. The other half had no special intervention. After five years, blood pressure was controlled in 63 percent of those in the intervention group, compared with only 22 percent in the control group.[26] Home blood pressure monitoring is another innovation that seems to encourage patients to continue to take their pills.

The need to identify and control hypertension remains urgent. We've long known that high blood pressure is a common problem, but even those of us who spend our lives researching its consequences were surprised by results of a Framingham paper published in *JAMA* in 2002.[27] The lifetime risk of developing hypertension, never before quantified, is 90 percent. It is, we found, a near certainty. This knowledge has placed greater emphasis on the transition from normal blood pressure to hypertension. Framingham findings were critical in motivating national guidelines to establish a new blood pressure category we now call "prehypertension," or a reading of 120 to 139 over 80 to 89.

We cannot simply wait for blood pressure to rise to health-threatening levels, and then prescribe drugs. As Jeremiah Stamler puts it, that approach is "late, defensive, reactive, time consuming, associated with side effects, costly, only partially successful, and endless."[28] Given our current lifestyle, diet, and genetic makeup, we are almost all destined to develop this disease. Waiting for it to strike and then using drugs to lower it is, itself, a high risk approach. Much wiser is to find the problem early and intervene through lifestyle modification, salt restriction, and a diet rich in fruits, vegetables, and nonfat dairy products, low in saturated fat, and high in fiber, potassium, and calcium. For many, those early changes, along with weight reduction and moderation of alcohol consumption, can reduce blood pressure as well as drug treatment. The knowledge that blood pressure creeps up as we get older should make us more vigilant about checking and intervening earlier. A half century of cooperative scientific efforts has given us an arsenal of information and an array of choices beyond anything Freis could have imagined.

He still thinks of the men who volunteered for the VA trial. He knows what they looked like, knows they had been to war, knows they all wanted to get better. They were not high-powered men, sitting around corporate conference tables. They were working-class individuals who had no idea they were participating in a project that would improve lives for generations to come. They knew only that they were part of a study to test treatment of high blood pressure. Freis stresses the uncertainties of what the VA trial would reveal:

> You have got to realize the justification for the placebo group was that most people didn't believe treatment did any good. There were plenty of guys using no treatment. So we didn't consider it a great crime to not treat some of the men. We decided it was necessary because we were not absolutely sure that this treatment did any good.
>
> Hindsight is always easy. People could say, "What in the hell were they doing letting these people die?" I did have a few doctors write to me, very mad about placebos. They were among the very small crowd that already believed in treatment. But what about the rest of them, doctors who weren't treating? What about what was happening before the study? What if we hadn't done it?

ELEVEN

The Wages of Sin

By the time Ed Freis published his groundbreaking results on hypertension treatment, other scientists were focusing their attention on the role of cholesterol in heart disease. Uncovering the deadly potential of cholesterol was an equally important discovery that helped establish new means of preventing heart disease. Even before the Framingham participants allowed themselves to be studied, laboratory scientists knew that the plaque buildup characteristic of atherosclerosis contained crystals rich in cholesterol. By the end of the 1940s, a scientist at the University of California at Berkeley, Dr. John Gofman, developed a method of separating blood into different components, cholesterol being one of them. At the Donner Laboratory there, he enlisted several graduate students to build an analytical ultracentrifuge for that purpose. Gofman and his coworkers ran the analytical ultracentrifuges to develop a new system of analyzing lipoproteins.[1] This approach allowed the identification of what we now recognize as "bad" cholesterol, or LDL (low-density lipoprotein), and "good" cholesterol, or HDL (high-density lipoprotein).

As early as the 1950s, Framingham researchers sent blood samples to Gofman's lab for lipid density measurement in his ultracentrifuge; that lab was then the only place on earth equipped to separate and measure blood in such a way. By 1957, Framingham researchers alerted the scientific world to the dangers of high total cholesterol. Still, well into the 1980s doctors thought that only total cholesterol levels higher than 300 milligrams per deciliter (mg/dl) needed treatment. The Heart Study would continue to parse cholesterol components, refining the alert. The number that would raise a red flag dropped lower and lower until today, when people are urged to modify their diet and exercise habits if total cholesterol exceeds 200. But epidemiology was merely one branch of science, and not the branch responsible for developing and testing treatments. Already, drug companies were looking into pharmacological inhibition as a tool for eventually controlling cholesterol in human beings.

The knowledge that cholesterol could be found in atherosclerotic plaque came from animal and human autopsy studies. Inspired by the work of Nicolai Anitschkow, Mark Armstrong and William Connor started experimenting on animals. They pureed the typical American diet—steak, mashed potatoes, and gravy—and added banana meal for familiar flavor. Their aim was to induce coronary artery disease. They fed it to monkeys, and the animals promptly developed atherosclerosis. Years later, they would prove that when the monkeys were allowed to return to their natural diet, their atherosclerosis would regress. Other laboratory researchers were feeding beef or cheese or straight fat or pure cholesterol to chickens, rabbits, rats, dogs, monkeys, and cats. Many of those animals developed plaques that blocked their arteries. Scientists showed that blood cholesterol in an animal could be taken from a normal level of 50 mg/dl to, in some cases, as high as 2,000 mg/dl by a change in diet.[2]

Autopsy studies of soldiers killed in the Korean War, and later in Vietnam, added even more clues. In 1953 in *JAMA,* Major William Enos described the autopsy evidence of diseased coronary arteries of young, healthy U.S. soldiers killed in action in Korea.[3] What he found was shocking. Although their average age was only twenty-two years, over three-quarters had some evidence of coronary plaque buildup, and 15 percent had a major coronary artery that was blocked by 50 percent or more. These findings offered science a startling view of how early these diseases began, and how common they were.

Epidemiological studies, including Framingham, pointed to evidence of an increased cardiovascular risk as cholesterol levels rose. High levels of cholesterol were known to cluster within families, and decades before the mapping of the human genome, scientists knew of a genetic disorder, familial hypercholesterolemia, in which cholesterol levels were twice as high as normal. The disorder often led to heart attacks in young adults. In rare situations in which someone has inherited two copies of the defective familial hypercholesterolemia gene—and only about one in a million does—cholesterol levels are above 1,000 mg/dl, and the person can have a heart attack as young as the age of five.

With these clues in hand, two men who met in 1966 in the early years of their medical training at Massachusetts General Hospital revolutionized the treatment of cardiovascular disease. Michael Brown and Joseph Goldstein used radioactively labeled LDL, "bad cholesterol," to find that our cell surfaces have receptors that take LDL cholesterol out of the blood. Without enough LDL receptors, LDL cholesterol is not sufficiently removed from the blood, and so it can build up as plaque, eventually clogging vessels and causing heart attacks and strokes.

Before science could develop powerful new drugs to treat high cholesterol levels, researchers had to understand the mechanism of how the body controls LDL cholesterol. Brown and Goldstein moved to the National Institutes of Health in Bethesda in 1968, learning sophisticated biochemical techniques and looking for a disease to study that would benefit from their new skills. Their lab was next door to Dr. Donald Fredrickson, former director of the National Heart Institute, who was studying patients with familial hypercholesterolemia. Fredrickson's group was treating two patients at the time, siblings aged six and eight, whose blood levels of cholesterol were 1,000 mg/dl. "These two children were hospitalized there because they were having multiple heart attacks. They couldn't walk across the room without an episode of severe angina," Brown recalls.[4] He and Goldstein didn't treat these youngsters directly. And the physicians who struggled with them had precious little to work with. "There was almost nothing you could do for them. Coronary bypass was not invented. There was no possibility for heart transplant," says Brown.

The field they were about to revolutionize, cholesterol research—indeed, heart disease research in general—was wide open. The grant money, the prestige, and the scientific clamor at that time were largely directed at cancer research. "It was a negative thing to work on heart disease. It was not very fashionable. Most of the bright young scientists in the late sixties and early seventies were interested in cancer or in brain degeneration. There was a relatively small number of people working on cholesterol," says Brown. "That was very attractive to us. It was nice to be in a field where you didn't have to worry that someone was going to beat you."

They worked in relative isolation, in their laboratory, with cells from the unfortunate youngsters who died in childhood. They compared the children's cells in tissue culture with cells from patients with normal levels of cholesterol. They found that normal cells were able to bind LDL, internalize it, and break it down. "Then we were able to show that the ability of normal cells to take up LDL depended on having a certain protein on the surface of the cell called the LDL receptor," says Brown. The LDL particle binds to the receptor on the cell surface, allowing it to be removed from the blood, internalized into the cell, and broken down. More receptors mean lower LDL levels, fewer receptors higher levels.

Ultimately, they created a mouse model with a genetic disorder similar to the children's. They studied more cell cultures, figured out how the liver processed cholesterol, and how it shut down receptors when it got overloaded, leaving overabundant LDL to flow through arteries, deposit and accumulate in artery walls, and build up as life-threatening plaque.

Brown explained it in a lecture in Washington, D.C.:

When cholesterol gets into the intestine, it goes to the liver and doesn't affect blood cholesterol. But with time, the level of cholesterol in the liver builds up. There's no way for the liver to say to your brain, "I have had enough cholesterol. Don't eat any more." There's no feedback mechanism, like "I'm not thirsty anymore, I don't need any more water." With no feedback mechanism to regulate the intake of cholesterol, the liver says, "I can't stop you from eating cholesterol, but I can cut down on LDL receptors." The liver regulates the number of LDL receptors, and over time, the production of LDL receptors is reduced.[5]

LDL particles deliver cholesterol from the liver to tissues that need it, but even very small amounts of LDL are sufficient to meet these needs. With too much LDL cholesterol, the liver is overwhelmed, too much is delivered to arterial cells, and it builds up as atherosclerotic plaques. With that evidence in hand, and as organ transplant techniques improved, surgeons theorized that liver transplants in patients like the young siblings treated at the National Institutes of Health in 1968 would result in an almost immediate drop in levels of LDL. "So we knew that the trick in controlling the level of LDL in the plasma was to increase the number of [LDL] receptors in the liver," says Brown.

Other cholesterol research was occurring in numerous labs. When Brown and Goldstein began their work, there was already a drug on the market that could lower cholesterol, though not nearly enough to help patients with familial hypercholesterolemia. And like the early treatments for high blood pressure, the treatment for elevated cholesterol levels seemed worse to most patients than the painless, invisible symptoms. In the 1970s, the NIH began the Coronary Primary Prevention Trial in several centers across the country, testing the theory that lowering cholesterol levels would reduce heart attacks. Scientists used a drug called cholestyramine, a resin that could decrease cholesterol levels by 10 to 20 percent. The trials showed that patients given the drug not only had lower cholesterol levels, but also experienced fewer heart attacks. It was the most convincing evidence at the time that diminishing cholesterol lessened the risk of heart attacks.

The drug came in sachetlike packets of gritty resin. Patients had to take eighteen or more grams of the sandy powder each day, and it caused bloating and constipation. Many couldn't stay on the treatment in the recommended dose. But from then on, scientists had the evidence linking high LDL cholesterol level and heart disease risk and the proof that disease was prevented by lowering LDL.

There are more than two dozen steps in the cholesterol biosynthesis process. Blocking synthesis too early could be ineffective. Interrupting it too late could have deleterious consequences. There was a lot more work to be done, and researchers interested in cholesterol drew on an earlier breakthrough from 1958 when scientists at the Max Planck Institute in Berlin found the major rate-limiting step in cholesterol synthesis: 3-hydroxy-3-methylglutaryl-coenzyme A, or HMG-CoA.[6] It was the enzyme that Brown and Goldstein proved important in the production of excess LDL. They reported in 1973 that oxygenated sterols reduced the activity of HMG-CoA reductase in cultured cells.

That information gave Akira Endo of Japan the tool he needed to discover the first statin. He made an extract of a penicillin mold and found in that extract a compound called compactin that was able to block the production of cholesterol by the liver and lower levels of cholesterol in the blood.[7]

Endo's finding would prove to be both a boon and a setback for development of a truly effective pharmaceutical therapy to treat hypercholesterolemia. The boon came first, when the Japanese work helped Brown and Goldstein, who were consulting with Merck on the development of such a drug. "We realized that if you could block the production of cholesterol by the liver, that would lower the cholesterol in the liver, and then the liver would turn on the [LDL] receptor gene," Brown explains. Scientists at Merck were working on what would be the first statin approved by the U.S. Food and Drug Administration. Because the FDA accepted the idea that lowering cholesterol would reduce the risk of heart attacks, pharmaceutical researchers would have to show only that their drug cut down on cholesterol.

Industry was ready to create a drug that was both effective and well tolerated. But then the Japanese research came to a mysterious halt. The effort behind Endo's drug, compactin, was suddenly abandoned. No paper was ever published on why, but scientists have speculated that the researchers found some serious animal toxicity, most likely a form of cancer. And the unofficial, unpublished word whispered around labs was "lymphosarcoma."

The fear of some unknown cancer-causing mechanism slowed down the entire industry. Merck had been working on its own statin, lovastatin. Jonathan Tobert, who was part of the effort, describes the Merck reversal: "We had demonstrated that our compound worked very effectively. Then in September 1980—I'll never forget—I got a call from the vice president of my area. He was obviously in a grim mood. He said we have to stop, because a very closely structured compound was rumored to have this very serious toxicity."[8] No one was even certain which animals were used in the experiments in Japan, nor exactly what the problem was. But the rumors were enough to set back the research by more than three years. Broader clinical trials testing statins on human beings, barely begun in 1980, were abruptly halted in 1981. "I made the decision to discontinue clinical trials of lovastatin because of rumors (to this day never substantiated) that the closely related compound, compactin, caused certain cancers in dogs. Nothing we had seen with lovastatin had given us any cause for concern, but we could not ignore the rumors about a chemically related HMG-CoA reductase inhibitor. It appeared that the lovastatin project was dead," wrote P. Roy Vagelos, chairman and chief executive officer of Merck, in the May 24, 1991, issue of *Science*. Merck scientists went back to animal research in the lab.

A rule of thumb in drug research is that what works in a laboratory rodent may not work in a human being. But it's also true that what is toxic to a rat may not be toxic to a dog, or to a human being. The ultimate proof comes in human clinical trials. One example of the drastically different effects among animals is that penicillin kills guinea pigs. Even chocolate can kill a dog.

The FDA had granted approval for compassionate use of lovastatin for patients with hypercholesterolemia who had no other hope for recovery. It had already been tried successfully on a few patients who were severely ill.

Stormie Jones of Dallas, Texas, was one of those patients at the age of six. She almost certainly would not have seen her seventh birthday were it not for the work of Brown and Goldstein, introducing an experimental drug with rumored carcinogenic effects, combined with a risky and experimental surgery.

Stormie was a genetic one-in-a-million child who inherited two faulty copies of the gene for familial hypercholesterolemia, and in her young life had already suffered a heart attack, two bypass operations, and a heart valve replacement. She was born without receptors for low-density lipoprotein and therefore had no way to regulate her cholesterol level. Her HMG-CoA was churning out cholesterol nonstop, but she had no receptors to remove it from her blood, and so her cholesterol blood levels were elevated to more than ten times the normal levels. A heart transplant would have solved the immediate problem, but with a liver devoid of LDL receptors, her new heart would have fallen prey to clogged arteries, too. Brown and Goldstein realized that she needed a new liver as well. And in 1984, when a four-year-old girl in New York died in a traffic accident, Stormie got her chance at life. Her double transplant, the first performed on a child, was historic, and proved a success when her cholesterol plummeted to near normal levels shortly after the operation.[9]

At just the right moment, the first HMG-CoA reductase inhibitor, lovastatin, became available to her for compassionate use.[10] So with the help of the first statin, her circulating cholesterol fell to normal levels. The success of the treatment became a medical precedent that would help millions.

Stormie lived for five years with few medical complications. She became bored with people asking if she was okay. She felt like a normal child, and wanted to be treated like one. But during her last year, she suffered hepatitis, and needed a second liver transplant. She died at thirteen, an unwitting medical pioneer.

The still unapproved drug was administered to a select few patients because without something new and experimental, they would have died. Roger Illingworth of Oregon Health Sciences University and Scott Grundy and David Bilheimer of the University of Texas used it to treat patients with severe hypercholesterolemia.[11] They found that it dramatically lowered LDL and total cholesterol in the blood, and had few side effects. Meanwhile, Merck scientists continued their laboratory research using lovastatin with great experimental success in dogs. Still, fearful of toxic side effects, the industry considered developing the first U.S. statin as an "orphan drug." The federal government, in order to encourage pharmaceutical companies to produce drugs for rare illnesses, provides certain tax and marketing incentives to companies to find drugs to treat diseases that affect fewer than 200,000 patients nationwide. In 1982, some at Merck and the FDA, still wary of an unknown toxicity, believed statins might be useful only for those with familial hypercholesterolemia. "This is probably the only drug ever considered for orphan development that, within four years, became a billion-dollar drug," says Tobert. Clinical trials testing lovastatin in patients at high risk for coronary disease began again in 1984, and results were apparent in months. Nothing before had ever so drastically lowered cholesterol levels in the blood. Meanwhile, years of very-high-dose treatment with lovastatin in dogs proved safe—no tumors or other serious side effects. Merck went to the FDA in October 1986 with 160 volumes of data from test tube and animal studies, and information on 1,200 people in clinical trials. Lovastatin, a breakthrough drug, was approved in August 1987 for patients with high cholesterol levels that could not be reduced sufficiently by diet.[12]

One by one, studies proved that a variety of statins lowered cholesterol and saved lives. A turning point in accepting their value came in 1994 when a study of 4,444 patients, called the Scandinavian Simvastatin Survival Study, showed an overall reduction in risk of 37 percent in patients with coronary heart disease who took the statin compared with those on a placebo.[13] Commonly called the 4S study, it proved that lowering cholesterol in patients who had already suffered a heart attack prevented a second one. It offered the first definitive proof that statins could lengthen the lives of people with heart disease. And it provided the reassuring result that individuals who took statins were no more likely to die of cancer—or suicide or violent or accidental death—than those on a placebo.

In a study of nearly 6,600 men with high cholesterol levels (average of 272 mg/dl) but no history of heart disease, called the West of Scotland Coronary Prevention Study, investigators randomized participants to receive either placebo or the drug pravastatin.[14] Fewer men assigned to the statin drug died from cardiovascular disease compared with those given a placebo. The 32 percent reduction in cardiovascular deaths extended our knowledge of the benefits of lowering cholesterol in people with elevated cholesterol but no prior history of coronary problems.

As encouraging results came in, researchers began testing statins on a broader range of patients, even on those with relatively normal levels of cholesterol. The Air Force/Texas Coronary Atherosclerosis Prevention Study showed that lowering cholesterol levels benefited healthy men and women who had average—not high—levels of cholesterol.[15] This study had a new wrinkle; it enrolled people who were predicted to be at increased risk for heart disease because of below-average levels of HDL cholesterol. As in other statin trials, heart attack rates were reduced by more than one-third in those who were given lovastatin compared with the placebo group. We learned from this study that even people with what we call "normal" cholesterol, if they have a low HDL level, can benefit from statin use.

In the largest cholesterol-lowering study ever carried out, British researchers gave simvastatin or placebo to 20,000 men and women at high risk for heart disease.[16] Deaths from cardiovascular disease were reduced by 17 percent in those taking the statin, and strokes fell by 27 percent. The benefits of lowering cholesterol were found to be similar in men and women and in younger and older participants. Importantly, those benefits were similar in patients with lower cholesterol levels prior to enrollment and in those with the highest levels. There was no evidence that patients with heart disease can lower their cholesterol too much. And new studies add further evidence that lower is better.

Still, the statins are relatively new, though lovastatin is now off patent and prices have come down. No one yet knows what effect decades of taking the drugs will have, though so far they appear remarkably safe.

Brown understands that his work has changed the practice of medicine. Giving statins to healthy people with normal cholesterol was not without its nightmare scenarios as the treatment has spread to millions. "Sure, there were fears. Once a drug goes into human beings, you can never predict exactly what's going to happen. Even if you do trials in five thousand human beings, you always hold your breath when it's suddenly being taken by millions. Right now there are 15 million people taking statins. There was always the chance that some rare form of toxicity would appear. And there's still such a chance. So far the drugs look really quite benign. But it's always a continuing concern," he says. Today, Americans have a choice of fifteen cholesterol-lowering drugs, with nine more in the research pipeline.

Sharyn Weir is one of 15 million now using a statin. When she was forty-two, her total cholesterol was 567.[17] She had been taking medicine to control high blood pressure since the age of nineteen. Family history was against her. Her father developed coronary symptoms beginning at the age of thirty-four, and died years later of a massive myocardial infarction. Her brother has had bypass surgery. Her mother had a heart attack at sixty-five. When Weir was a young woman, even with a terrible family medical profile, her doctor didn't put much stock in cholesterol numbers. He simply didn't think women her age had too much to worry about from a high cholesterol level.

A prime-of-life heart attack at the age of forty-two is not something you forget. "It felt like an elephant sat on my chest, and the pain went up to my jaw. My whole left arm went numb. Afterward, it was very frightening. I was afraid to walk up and down the stairs. Every little twinge I got, I thought it was another heart attack."

At the time of her heart attack, there were effective cholesterol-lowering drugs, but she had not been given a prescription. Even now, fewer than 25 percent of those who probably would benefit from cholesterol-lowering medications take them. She tried to improve her diet but made little impact on her numbers. "I never exercised. I never had to. I weighed 110 pounds before my heart attack," she says. "The attack was a wake-up call."

With no apparent symptoms, it took a medical disaster to get her to change. She was afraid that life would never be the same, but today, the odds are with her. She's a second-generation volunteer in the Framingham Heart Study, and after her illness she started attending William Castelli's prevention clinic. She learned how to get her numbers under control. She joined a gym, but never liked public exercise. She bought an exercise bike and treadmill, and used one or the other every day. Eventually, she found tai-bo, an exercise form that's a combination of tai chi and kick boxing, and fell in love with it.

Now, her total cholesterol is 128, her HDL is 39, and her LDL is 70. She does it largely with lipid-lowering drugs. She has severe hypercholesterolemia and cannot rely solely on a healthy diet and lifestyle. For Weir and many others like her, medication is crucial.

She eats oatmeal every morning; snacks on apples, oranges, and bananas; has tuna or turkey (without the mayo) on whole wheat for lunch. Her banana muffins are now made with applesauce instead of butter or shortening. She shuns processed foods and eats fish or chicken (without the skin) for dinner and a lot of beans, vegetables, and salads.

She has benefited from the best that medicine and nutrition research have to offer, even though initially it was hard for her to accept that she couldn't make a dent in her lipid profile without prescription drugs. "What I have is inherited, and there's no way I can do it naturally. I had to come to the understanding that the drugs are helping me," she says. "And I feel wonderful. It's not true that if it's in your genes you can't do anything about it. You have to work harder at it. And I have two beautiful grandsons that keep me ticking."

It's easy to say what the "average" cholesterol level is in the United States—about 210. It's harder to say what "natural" cholesterol levels would be without junk food, fast food, and high-fat diets. "That number keeps changing," according to Brown. "In other animal species, including primates, the level of LDL is much lower than in Americans. And if you look at newborn infants, the LDL level is below 100. If you grow up in China or Japan, your cholesterol will stay lower throughout your life. In rural China and Japan, the rate of heart attack is twenty-five-fold lower than in the United States. And if you grow up in indus-trialized countries, cholesterol continues to rise with age. We tend to think the Chinese levels are normal. That would be a total of about 160."

Americans clearly have a long way to go, and a study of 4,162 with heart disease indicates that perhaps they need to go even further than previously thought. In a 2004 report in the *New England Journal of Medicine,* researchers found that patients who took a high dose of a statin (atorvastatin) lowered their LDL levels from 106 to 62. Those on a conventional dose of another statin (pravastatin) saw their lipid levels drop, too, but only to 95. More importantly, the aggressively treated group had 16 percent fewer deaths, heart attacks, strokes, or other car-diovascular events than those who were treated conventionally.[18]

I have long heard Castelli talk about the peasants of rural China and the fact that they have exceptionally low LDL cholesterol levels—similar to what was achieved with aggressive treatment in this study—and they don't develop heart disease. I always thought it was his dramatic way of making a point, but one that would never be tested. This clinical trial proved him right. Very low is better than low and what was previously thought to be an acceptable goal of treatment—an LDL cholesterol level less than 100—is not as protective as a level less than 70. This groundbreaking study would have been unthinkable prior to the discovery of statins because it would have been virtually impossible to achieve such low levels of cholesterol. It's still unclear what the most desirable LDL level ought to be for people who do not yet have heart disease, but the recommended level will no doubt drop further.

For their extraordinary body of research on cholesterol, Michael Brown and Joseph Goldstein were honored with the Nobel Prize in Medicine in 1985. And for her contribution, Stormie Jones has a park named in her honor in Dallas, Texas.

TWELVE

Renegade on the Trail of the Unknown

I had been working at the Framingham Heart Study for several years before I learned about homocysteine as a potential risk factor for heart disease. It's not an idea that originated in Framingham, but it's another illustration of how medical research is shared. Hypertension and high levels of LDL cholesterol in the blood were found to be risk factors in Framingham Study participants. Later, other scientists proved through clinical trials that their modification can reduce the risk of cardiovascular diseases. Homocysteine was investigated at Framingham only after the curiosity of one of our researchers was piqued by a twenty-year-old theory on which a maverick researcher in a laboratory had staked his entire career.

In 1995 a Heart Study publication, in collaboration with Jacob Selhub and Paul Jacques of the Human Nutrition Research Center at Tufts University, was one of the first studies to support Kilmer McCully's life-work.[1] By then, the Framingham participants had been watched for more than forty years, but we had measured homocysteine levels only recently. "It was the first and best study of the dietary intake of B vitamins," according to McCully.[2] His interaction with Framingham researchers began indirectly when a young physician he knew in Rhode Island, Andrew Bostom, expressed interest in the causes of vascular disease. Hours of discussions with McCully ignited an interest in homocysteine. And when Bostom then went to work at the Framingham Heart Study as a research fellow, he brought that passion with him. Bostom began working with nutrition scientists at Tufts, and that university and the Heart Study formed a partnership to consider homocysteine as a risk factor. "We incorporated it into the exams. We have been

:eptive to new variables as they appear in the scientific commu-
have the population to do this," says William Kannel.

ien Selhub and his group examined 418 men and 625 women
Heart Study and found a relationship between dangerous ob-
structions of the carotid artery and blood level of homocysteine, it
strongly implicated high levels of homocysteine as a risk factor for vascular disease. And disease risk rose in tandem with homocysteine levels.

The Framingham result, set in motion by McCully's work, came
back to further encourage him. "That finding made a huge impression
on me and everyone else," says McCully.[3]

The epidemiological numbers matched up—30 percent of Americans were not getting enough folic acid, and 30 percent of Framingham
volunteers had high homocysteine levels. At about the same time,
results from the Physicians Health Study, a long-term study of 15,000
physicians in the United States, found that homocysteine levels correlated with cardiovascular disease risk.[4] Other large-scale studies followed suit. The Scottish Heart Health Study and the U.S. Nurses
Health Study in 1998 found a relationship between homocysteine levels
and cardiovascular disease in Scottish men and women and in American
middle-aged women, respectively.

The burden of scientific evidence from these studies finally put
homocysteine on the map as a potentially important and modifiable risk
factor. As with blood pressure and LDL cholesterol, the definitive
answer ultimately will come from clinical trials of homocysteine-
lowering therapy and the cardiovascular benefits. And clinical trials to
test the hypothesis would be relatively inexpensive to conduct. After all,
lowering homocysteine is as simple as taking over-the-counter vitamin B6, vitamin B12, and folic acid, which cost only a few pennies a day.
Those studies are now beginning to weigh in on a story that actually
began one hundred years ago.

It's a tale that has deep personal roots for Kilmer McCully. For him,
it goes back to the immigration of his grandparents to the wheatlands of
the upper Midwest.[5]

In 1888, riding a horse and buggy, Judah Litwinenco, his wife
Marie, and their three children arrived at the end of the railroad line in
Aberdeen, in the Dakota Territory. They saw no house, no fence, no
town, scarcely a tree, and certainly nothing like a power line. It was a
raw, undeveloped, but fertile paradise of chest-high prairie grass rip-

pling in the wind. Midwesterners describe the endless sky and wind-blown fields in much the same way that coastal residents talk about the ocean: waves pulsating to the rhythms of nature. Back then, it was a sea of prairie grass, but later it became fields of fine durum wheat.

When the Litwinenco family first saw this land, it was only a decade after Chief Sitting Bull and his warriors were defeated by the U.S. Army and forced west of the Missouri River to an arid plain that became a Sioux reservation. The fertile land east of the Missouri River was open to white settlers, who could receive title after ten years of farming.

Judah Litwinenco was the first white settler on this land, which became South Dakota. He bought the horse and wagon with savings earned as a hod carrier in New York City. It's one of those backbreaking jobs that no longer exists, though most Americans can conjure an image of the V-shaped wooden troughs in which workmen once carried the bricks that built cities. Judah was a big man with a round face, carrying about 210 pounds on his five-foot-eleven frame. His hands were massive and strong. "When you shook hands with him, it was like shaking hands with a side of beef," says his grandson, Kilmer McCully.

Judah built a sod house for his family, sod being as abundant as trees were scarce. He put to use the seeds he had gathered from his farm near Odessa that he had carried in his pocket across the Black Sea, the Atlantic Ocean, and half the North American continent. He had smuggled them past customs inspectors and guarded them ever since he left Ukraine.

In this way, Litwinenco was among the first pioneers to cultivate durum wheat in the American wilderness. The wheat was planted in the spring, grew through the early summer, and, at two to three feet high, was ready to be harvested in August. Judah and Marie stone-ground it and used the flour to make bread.

It is, according to their grandson, the finest wheat grown anywhere. In its pure, unrefined form, it is an excellent source of the B vitamins that may protect against heart disease by limiting the amount of homocysteine circulating in the blood. A quest to prove that benefit would define, and nearly ruin, McCully's professional life nearly a century after his grandfather cleared and plowed the wilds of South Dakota.

McCully tells the story of his grandfather in a 2001 journal article that illustrates an unusual connection that spans three generations.

McCully grappled with the issue of homocysteine as a risk factor for heart disease from the late 1960s well into the twenty-first century. To the medical establishment of the time, he must have seemed like a hapless searcher on the trail of a culprit, in a pursuit they no doubt dismissed as either innocent or irrelevant. For his ideas he went through decades of ridicule and humiliation. His career stagnated, then stalled. There will always linger a hint of *what if.*

In 1968, McCully was a newly appointed Harvard pathologist, listening to a lecture about a rare condition at a human genetics conference. The lecturer recounted the story of the 1932 death of an eight-year-old boy. The long-forgotten case had been published in the *New England Journal of Medicine* in 1933. More than thirty-five years later, the lecturer mentioned this case of a boy who died at Massachusetts General Hospital just as an old man might have died, of a massive stroke. McCully had also heard a very recent report about a nine-year-old girl who had homocystinuria, a disease newly discovered and named only three years earlier, in 1965. She turned out to be the niece of the boy described in the lecture who died in the 1930s. The girl's mother remembered the long-ago tragedy of the premature stroke in her brother, and told pediatricians. A contemporary physician studied the earlier case, and, putting new research together with old clues, diagnosed the boy with homocystinuria thirty-six years after his death.

The presentation caught McCully's attention. "I restudied that old case from 1933," he recalls. "But I didn't know quite what to do with it. It wasn't as if a lightbulb went on at the moment. Because I knew something about homocysteine, I began to read all the articles that were available on the subject. And the more I read, the more interesting it got."

The boy's death was mysterious at the time he died. Pathologist Tracy Mallory concluded that he died of arteriosclerosis of the carotid arteries, the arteries clogged by a "process one might expect to find in a very elderly man." Mallory published the results in 1933. The underlying cause of death was inexplicable, and unusual enough to have left a record of autopsy reports and preserved tissue samples.

The year the child died was the same year that Vincent DuVigneaud discovered a new amino acid, which he named homocysteine.[6] Throughout the 1940s and 1950s, little was learned about homocysteine, and nothing was known about its significance in medicine. In 1962, researchers in Belfast, Northern Ireland, used newly developed techniques to screen the urine of children with mental retardation, looking for abnormal amino acid levels.[7] Homocysteine is present in small amounts in everyone's blood, but in some mentally retarded children it was found in large amounts in the blood and urine. The disease was named homocystinuria.

When McCully heard of these cases, no one had yet begun to link the clues to arteriosclerosis. He began delving into the files within the hospital for details of the case of the eight-year-old boy. The original autopsy report said that he suddenly developed a headache, drowsiness, vomiting, and left-sided facial weakness. Despite medical treatment, his condition quickly deteriorated and he slipped into a coma and died. McCully was able to find samples of some of the boy's organs, preserved in paraffin blocks. He examined the tissue and was shocked to see under the microscope fibrous arteriosclerotic plaques scattered through the small arteries of many organs. There were blockages that might typically appear in an eighty-year-old, with a crucial exception. They had no buildup of fat or cholesterol. McCully theorized that in this newly named disease, the damage to arteries is severe and rapid, and can cause death even before the fat and cholesterol deposits associated with coronary artery disease appear.

Four months later, while McCully was sleuthing in the hospital's medical archives, reading everything he could on homocysteine and examining slides made from the boy's pathology specimens, he heard of another case that had recently come to Massachusetts General Hospital. It was a two-month-old baby, and this time, medical science knew enough to diagnose the infant's disease as homocystinuria, though it was not the same type as the case of the eight-year-old. The contemporary one resulted from a lack of vitamin B6, and the older case from a lack of vitamin B12 and folic acid. McCully thought back to the research literature. In 1949, James Rinehard deprived laboratory monkeys of vitamin B6 and created arteriosclerosis in the primates. McCully searched the medical literature and came upon other experiments showing that giving B12 and folic acid to rats prevented arteriosclerosis. The link, he postulated, was homocysteine, raised in the blood by insufficient levels of B6, B12, and folic acid. Perhaps, if very high levels of homocysteine caused homocystinuria, then more modestly elevated levels of homocysteine could damage the arteries in otherwise healthy people and contribute to atherosclerosis. This led McCully to wonder if a diet rich in B vitamins and folic acid could help prevent heart disease not only in those with the rare disorder, not only in rats and monkeys, but in normal, healthy humans.

Despite treatment, the baby died of bronchopneumonia. The autopsy results were on file in the pathology department, and McCully looked for mention of changes in the arteries. There was none. "I realized how important this case was," he says. "I realized if the child had no lesions in the arteries, then my theory was wrong, there was nothing to it. But if they were damaged, if you had high homocysteine for any reason, then it had to be true."

This time, the baby's body was preserved in formaldehyde. McCully spent weeks examining tissue under his microscope, and detected rapidly progressive arteriosclerosis. "I was astounded to find that the baby had both early and astonishingly advanced arteriosclerosis plaques involving arteries throughout the body," McCully wrote in his 1995 monograph. "When I first saw the arteriosclerosis in the arteries of this two-month-old baby," he recalls, "I knew I was right."

He was so excited by this discovery that he couldn't sleep for weeks. "I would wake up in the middle of the night thinking about heart disease and arteriosclerosis."

He proposed his theory in 1969.[8] "The possible role of elevated concentrations of homocysteine or its derivatives in the pathogenesis of arteriosclerosis in individuals free of known enzyme deficiencies will be discussed and interpreted with particular reference to the findings in experimentally produced arteriosclerosis," he wrote. Too much homocysteine in the blood, caused by not getting enough B6, B12, and folic acid from the diet, or because of a genetic disorder, causes the arteries to be damaged and plaque to form. Those vitamins, once abundant when Americans ate fresh fruits and vegetables and goods baked and cooked from whole grains, were largely missing from the modern diet of white flour and canned goods, the processing of which destroys many of the nutrients.

He published his paper, and thought something significant would come of it. Research scientists asked for reprints, mailing in several hundred requests in the first month. In the field of heart disease research in 1969 that was a good response, but not as good when one considers the thousands of cardiovascular disease researchers looking for answers. And the bulk of those researchers were on the cholesterol trail. At Harvard Medical School and Massachusetts General Hospital, his theory about a link between homocysteine and atherosclerosis fell largely on deaf ears. In retrospect, silence was the kindest response he would receive for two decades.

"Cholesterol had become almost a religion. People saw this homocysteine idea as minor, and they wished it would go away," he says. By 1969 cholesterol had been identified as a major risk factor, but evidence that lowering cholesterol reduced heart disease risk was still years off. Today the theories about homocysteine are not competitive with what we know about cholesterol. Nor are they mutually exclusive. Homocysteine and cholesterol levels are linked by the American diet. High homocysteine levels damage arteries, causing lesions. Then, overabundant cholesterol comes through the bloodstream and gains entry into the damaged arterial lining. McCully saw cholesterol as a cofactor in the homocysteine problem. "Homocysteine causes the damage," he says, "and cholesterol gets the blame."

But when McCully began his work, few had time or money for his theory. He soldiered on, studying animals and tissue cultures, getting closer to unlocking the pathway from diet to blocked arteries. Homocysteine comes from the breakdown of an amino acid called methionine. He found that the homocysteine molecule is metabolized by enzymes that interact with B vitamins. He knew from his studies of the children who died of homocystinuria that those with the disease had genetic defects that disabled those enzymes. And he supposed that even healthy people, if deficient in B vitamins, could shut down the work of the enzymes, as well.

McCully maintained initial support for his theory from his mentor Benjamin Castleman, then chairman of the Department of Pathology at Mass General. He continued working with laboratory animals, publishing papers, and accumulating material that added more and more weight to his theory. Castleman retired and, under the Harvard umbrella, McCully published papers supporting his theory until about 1975.

Robert McClusky became the new chief of pathology, and he had little interest in homocysteine. "People were unwilling to accept any view other than that fats and cholesterol were the cause of heart disease," says McCully. McClusky told McCully that if he didn't get additional grant money, he would have to leave.

In the world of cardiovascular research, McCully's timing couldn't have been worse. Just when he wanted to pass the baton to clinicians, epidemiologists, and trialists, cardiovascular disease researchers were preoccupied by a growing body of data on cholesterol. In 1976, Australian researchers Bridget and David Wilcken published the first human study showing a possible connection between high levels of homocysteine and coronary heart disease, but the theory received little attention in America.

Extensive cholesterol research was bearing fruit and establishing cholesterol as a key modifiable heart-disease risk factor. The National Institutes of Health was pouring money into cholesterol research to prove or disprove its importance in heart disease. Homocysteine was far less well established and a much lower priority for research. Many prominent scientists had staked their careers on cholesterol as the central risk factor underlying cardiovascular disease. Pharmaceutical companies were developing drugs to lower cholesterol, and had a huge stake in the massive market for treatments. Indeed, today cholesterol-lowering drugs are a multibillion-dollar business. But no one had a financial stake in hearing about McCully's renegade theory because, if he was correct, treatment with diet and over-the-counter vitamins that cost only a few pennies a day would enrich no one. Many clinical trials that test new drugs are funded by the companies that hold patents on those compounds. Vitamin B_6, vitamin B_{12}, and folic acid are generic and not subject to patent.

Despite the respect he garnered from studying with some leaders in the field, including Konrad Bloch, the Nobel Prize winner for cholesterol biosynthesis, most of his peers saw McCully as pursuing a dead-end theory. McCully lost his lab. Real estate within crowded academic hospitals is at a premium, and space is tightly linked to status. He was banished to the basement, and his staff was cut. "I'd been on the fast track," he recalls ruefully.

> I was affiliated with Harvard for twenty-eight years—college student, Harvard Medical School. I interned at Mass General. After two years at the NIH, I was a research fellow at MGH. I was on the staff for eleven years. Then when Ben Castleman retired in 1975, the new chief took away my laboratory and put me in another part of the hospital. But it wasn't just geography. They put me in an empty laboratory in the basement of the Ivory Tower. They weren't listening. They weren't

interested. Unless you renew your grants, you're out. I had had a grant for five years, but I could no longer work under these circumstances.

The hospital's director told him he had failed to prove his theory. A former classmate attacked his ideas as nonsense and a hoax. He endured a year and a half of humiliation from colleagues. He was asked to leave.

He lectured on his theories. Following one talk, he got a phone call from the director of public affairs at Mass General. "He said that he never again wanted to hear the name of McCully or the word 'homocysteine' associated with Massachusetts General Hospital or with Harvard," McCully says.

He was professionally dismissed and close to being literally out on the street. "We thought about mortgaging the house and renting an apartment." Instead, he and his wife, Annina, borrowed money from her parents and kept the house. He remembers the dates of those months when he could not get work: from January 1, 1979, to April 1, 1981.

"I would go to different job opportunities I heard about. Maybe seven or eight actually invited me for interviews. It would seem promising, then all of a sudden it would stop. I'd hear nothing more. I went through this fifty-one times. I have it documented in my files. Later, I found out there were phone calls being made. My punishment was unemployment for twenty-seven months."

The worst of it was that McCully didn't know how long it would last. "I didn't know exactly who was doing me in, and I didn't know how to get out of my predicament. I was afraid I wouldn't be able to work anymore, and I'd have to give the whole thing up. My income was cut off. My wife had to go to work. I felt very bad for my family, that they had to suffer from my ostracism." His son and daughter, both of whom once considered medical school, turned away from the profession.

When he finally accepted that he was being blackballed, he called a lawyer. "My attorney, William Homans—he's now deceased—was a civil rights attorney. I talked to him for four hours. He said that he'd make a few phone calls."

Soon the trouble stopped. McCully was offered and accepted a position at the Veterans Administration Medical Center in Providence, Rhode Island. "That was in 1981. It was over."

He continued to work on his homocysteine theories in laboratory animals. He published a monograph, an overview of the history of the theory. He does not get grants but often pairs with other researchers doing work in the field. He has recently moved to the Veterans Administration Boston Health Care System, where he continues his work. At this point in his career he might have been head of a large lab, with millions of dollars in grant money carrying his research forward.

But it didn't happen that way. "I was dealing with big bureaucracies. I was a product of Harvard. I know what they're like. They have an extremely high opinion of their own abilities. There was some interest in my work, but for some reason the *anti* forces got the upper hand." But those forces never made him doubt the importance of his theory. "Not for a millisecond."

McCully says that as a lab researcher he never could have fully tested his theory. He worked with cell cultures and animals. Proving or disproving it would take collaboration with epidemiologists, and he watched us at Framingham studiously. It also required clinical trials—which, from McCully's perspective, came years too late.

The long-term epidemiological findings like those reported from the Framingham Heart Study in 1995 got attention, and human clinical trials followed. Eventually, 22 years, 10 months, and 29 days after McCully was ostracized, the *New England Journal of Medicine* published in its issue of November 29, 2001, these words: "Pyridoxine (vitamin B6) deficiency appears to be an independent predictor for coronary artery disease. . . ." Observations that total plasma homocysteine level is an important predictor of heart disease led to the question, Could lowering the level reduce the incidence of blockages re-forming after coronary angioplasty? Since homocysteine levels could be lowered 25 to 30 percent with vitamins, researchers hoped they could show that vitamin therapy could help keep arteries open after the medical procedure. In the study, 205 patients who had undergone coronary angioplasty to open blocked arteries were randomly assigned to two groups. One group received a placebo and the other group received treatment with homocysteine-lowering therapy: vitamin B6, vitamin B12, and folic acid. While angioplasty is a common, necessary, and successful procedure, it carries a high rate of restenosis, or subsequent blocking of the arteries. In this trial, those taking vitamin therapy saw their restenosis rate cut in half compared with the rate in those getting a placebo.[9]

More recently, a study published in 2004 reported negative results that would prove to be a blow to McCully's theory. In that trial 3,680 adults who had suffered a stroke were randomly divided into two groups, one receiving placebo and the other B vitamins and folic acid. Those receiving vitamins had a moderate decline in their homocysteine levels but no benefit where it really counts—reduction in occurrence of another stroke, a heart attack, or death.[10] Additional trial results are needed. It could well be that McCully will eventually be proved correct. Or it could be that he will be one of the thousands of researchers whose life's work benefits science through a rather unglamorous route: by showing a theory to be untrue.

Already some of the most staunch homocysteine proponents have begun to abandon the theory. I spoke with Bostom about his interpretation of the results of the recent clinical trials. He now believes that the theory about homocysteine may have been wrong. Homocysteine may be only a marker of disease activity and not a causal risk factor. The jury is still out. Ironically, key evidence may be difficult, if not impossible, to obtain. In order to reduce neural tube defects in babies, the grain supply in the United States is now fortified with folic acid, which will reduce high levels of homocysteine in the population—a late twentieth-century solution to a problem introduced in the late nineteenth century. This will interfere with the results of clinical trials of homocysteine-lowering therapy because everyone now eats enriched breads and cereals; even trial participants assigned to the placebo group will have their homocysteine levels lowered and will be difficult to differentiate from those taking B vitamins and folic acid. The jury may never reach a verdict.

McCully cannot help carrying some measure of regret at what was lost. "Most of my lab work has been collaborative. I couldn't get funded. With my history, nobody wants to give me money. I've not been able to get independent grant money. The only way I can do any work is to collaborate with people who already have money. My career would have been totally different if I had stayed at Harvard," he says. "I'm not at all bitter. Most critics involved are either retired or dead. They're not in a position of influence anymore. Younger people don't know or care. The only regret I feel is that if I had had funding and a better laboratory, I could have accomplished more. It slowed me down. It takes time to do research. It takes time.

"Wheat plays a role almost unique in human history," McCully says. Wheat is, metaphorically, in his blood. He knows its history, that it was once stone-ground into flour and made into dark bread. In the process, stone fragments, if not carefully sieved from the flour, could be left in the bread—too small to be seen, but big enough to chip teeth. In eighteenth-century America, many people had broken teeth. Then, late in the nineteenth century, along came a faster, more efficient, and more tooth-friendly way to process wheat kernels. The steel roller press released the starch granules to make white flour, and eliminated the wheat germ and bran. The white flour could be stored for months because most of the oils had been removed. With the stone-ground method, the flour had to be used quickly or it would go rancid. The by-products of the new process, wheat germ and bran rich in vitamins and nutrients, became cattle feed and pig slop.

In 1971, research by Henry Schroeder found that the way foods, including wheat, are processed in modern societies robs them of B vitamins and folic acid.[11] Canning also reduces vitamin content. By the time much food gets to American consumers—snack foods, fast foods, junk foods, convenience foods—it is almost devoid of the vitamins that regulate homocysteine levels. What had once kept our ancestors safe was being denied to most Americans. As more nutrition studies showed that the American diet included ever-decreasing amounts of B vitamins, the food industry responded. Companies still took out natural nutrients in processing, but they began to put synthetic nutrients back in. Most breakfast cereals today contain added B vitamins. The health food industry, a multibillion-dollar business, markets the supplementary addition of these once naturally abundant nutrients.

"It is ironic that the finest wheat in the world, the durum wheat of Dakota introduced by Judah [Litwinenco] and his fellow immigrants, loses much of its vital nutrient richness during milling by the roller process, leading to degenerative diseases in consumers of white flour," McCully has written.

Judah Litwinenco lived to be eighty-nine. He and Marie had ten children, many of whom lived to be 95 to 102. They had five grandchildren, four of whom became physicians. And one of these physicians, Kilmer McCully, contributed to a body of still unresolved research on the connection between nutrients found in durum wheat, vitamin B_6, vitamin B_{12}, and folic acid, and the prevention of vascular disease by lowering blood levels of homocysteine.

THIRTEEN

The Lifestyle Revolution

In many ways, the lifestyle battle has been harder on the sons and daughters of the original Framingham volunteers than on the volunteers themselves. Their offspring did not grow up baling hay, walking to school, or enduring wartime rationing. Faye DeSaulnier, says, "I watch my diet; I just eat too much. I know what I have to do. I just don't do it." Her mother, Marion Kittredge, now in her nineties, is a retired nurse who swims for exercise, and is in better shape than her children. She confides, "I don't know if my children will live to be as old as I am."

Framingham proved the Study founders right in being a typical American town—stable, consistent, and representative of how Americans live. But the eating habits of the participants, it turned out, were too typical, too consistent. It was not possible to illustrate how different dietary patterns affected the heart. With a few ethnic variations, Framingham residents ate the same foods. It was the American high-fat diet, heavy on meat and light on fruits and vegetables.

What the research world needed was a wider look, a global sampling of eating habits. Nutrition researchers like Ancel Keys and Jeremiah Stamler collected evidence that a diet high in saturated fats was connected to cardiovascular disease. But they had to go further to find the mitigating effect of monounsaturated fats like olive oil, and the value of fish over meat.

Keys was among the first to examine diet as a means of preventing heart disease. In 1959 he and his wife, Margaret, wrote *Eat Well and Stay Well,* a book that discussed the "pleasures of the table" and their health-threatening consequences.[1] He's one hundred years old now. In his prime, Keys was a bit stocky, but always energetic. His colleague, Henry Blackburn, has written that he "has a quick and brilliant mind, a prodigious energy and great perseverance. He can be frank to the point of bluntness and critical to the point of sharpness."[2] Keys had little patience with the overweight. "Disgusting" was how he referred to obesity. He has been quoted as saying that "Americans have Sunday dinner every day," and that heart disease was due to "the North American habit for making the stomach the garbage disposal unit for a long list of harmful foods." He disliked television because it encouraged sedentary life and is on record saying, "If we could find some way to make people do push-ups during commercials, then we'd all be strong as lions."[3] A respected scientist who knew what a high-fat, cholesterol-laden diet could do to a laboratory rat or a rhesus monkey, he felt that it was time to study the effects of an American diet on human beings.

His unique perspective on nutrition came from a lifetime of experience. In his early twenties, he sailed to China aboard the SS *President Wilson*. "The diet was mostly alcohol," he said in a 1961 interview. "I don't remember eating anything."[4] As a young man, he climbed a peak in the Chilean Andes, worked in a lumber camp, shoveled bat guano in an Arizona cave, served as a powder monkey in a Colorado gold mine, and clerked at Woolworth's.[5]

But he was always fascinated by diet and nutrition. He began his medical career commissioned by the government to study the effect on performance of nutritional deficiency. The thirty-six volunteers for the so-called starvation experiment were conscientious objectors. In the closing years of the war, he recruited volunteers to go without food for days, research that proved invaluable in rehabilitating millions of malnourished war victims. "Of course, we did a lot of screening because we found all the kooks in the world in that conscientious objector pot. But the Church of the Brethren and the Quakers made sure we had good, decent, honest youngsters who were not simply trying to stay out of the war but wanted to do some service," Keys said in 1979.[6] He is still in touch with some of the volunteers. Out of the work of those studies came the development of a minimally sound meal for combat, called the K-ration—the *K* for Keys. It contained bacon, canned cheese, dextrose tablets, gum, and cigarettes. The only thing thought to be important at the time was providing an adequate calorie count—with gum and cigarettes for relaxation. Keys bought the foods at a Twin Cities market called Witt's. He took his research to Fort Benning, Georgia, to develop the final research product.[7] At peak production, 105 million servings of K-rations went out to the troops in 1944. The content of the K-ration has changed, but not its name. Even now, what is minimally necessary to keep a combat soldier alive and functioning is hardly the stuff of a healthful diet.

It was the war, too, that launched him into the major work of his life, work that took place in a modest office on the campus of the University of Minnesota, under Gate 27 of the school's Memorial Stadium. It was under the stadium that he started what he called physiological hygiene, combining physiology, nutrition, epidemiology, and preventive research. One of the first patients to cross his path was a dairy farmer from Wisconsin who had knobs on his elbows and eyelids called xanthomas and xanthelasmas. The man clearly had familial hypercholesterolemia, an inherited disorder that speeds up atherosclerosis. Those with the disorder have extremely high levels of LDL cholesterol, and fat deposits form on their tendons and skin. When Keys cut open the knobs, they "were just pure cholesterol inside," he wrote.[8] The farmer's blood cholesterol level was an astonishing 1,000, but after an essentially fat-free diet, it came down first to 500, and eventually to 300, in the days before cholesterol-lowering drugs existed.

In 1951, on a trip to Naples, Italy, Keys and a physician friend measured the height, weight, and blood pressure of volunteers from the fire department. He also asked them questions about their diet and took blood samples. That got a community of researchers in Italy involved in diet studies. In Madrid, Keys took the same measurements on impoverished Spaniards. The firemen from Naples and the urban poor from Madrid all had lower cholesterol levels in their blood than the average American. One of the few things the Italians and Spaniards had in common was that they consumed less fat in their diets than did Americans.[9] He continued, with his wife, to study the calorie intake of fat in South Africa, among Bantus, Cape Coloreds, and Europeans. He found that fat intake varied widely among the groups, but that heart disease correlated with high animal fat intake. It was lowest among Bantus, highest among Europeans.

In the 1950s, with little to go on in nutritional research, it was clear that the Japanese ate a diet very low in animal fats, high in fish, and high in complex carbohydrates, especially the national staple, rice. So naturally, the Japanese were a target for Keys's curiosity. He studied them in their native land, then looked at them again in groups that had immigrated to Hawaii and Los Angeles. The more Americanized the Japanese diet became, the higher the risk of heart disease.

This research turned out to be background for his stellar achievement, the Seven Countries Study, funded in part by the NIH but with help from each individual nation involved. It started in 1958 to investigate the role of diet, fat, and cholesterol in cardiovascular disease by examining sample population pools from Japan, Yugoslavia, Finland, Italy, the Netherlands, Greece, and the United States. Keys and colleagues in each of the countries hypothesized that the occurrence of heart attacks and strokes—different in each of the populations—would bear some relation to the varied dietary habits within each country, especially to the amount and kind of fat in the diet. Researchers sampled diets from Japanese fishing villages and agricultural regions; from the east and west of Finland; from Dalmatia and inland Slovenia; from Corfu and Crete; from the Po Valley near the Adriatic in Italy; and in railroad workers in a Pullman car traveling across the United States. The men they studied were between the ages of forty and fifty-nine. They were farmers, fishermen, food processors, railroad workers, peasants, and professors.[10]

They were asked about their families, their work, what they ate, and how they lived. Scientists took a complete medical history. The men had a physical examination, had their body fat measured, received electrocardiograms, and underwent three-minute exercise tests. They gave blood and urine samples.

Their diets were measured by weighing all items consumed during a seven-day survey, which was repeated in different seasons. Nutritionists re-created the meals and estimated the nutritional content. Then the men were checked by an internist several times a year. Complete reexaminations were made every five years on 94.2 percent of the survivors.

But the researchers didn't just investigate the diet of others. They ate it themselves. Blackburn, who worked with Keys studying railroad workers in the United States and on elements of the Seven Countries

Study, wrote a memoir called *On the Trail of Heart Attacks in Seven Countries*.[11] In systematically examining the relations of diet and lifestyle to rates of heart attack and stroke in contrasting populations, Keys and his fellow researchers endured hardships and a lot of strange foods. Scientists in Yugoslavia were forced to use pledgets, or wads of dressing material, soaked in brine, to attach electrocardiograph electrodes because they had no electrode paste. Blackburn writes of getting a sore hand and wrist from stapling medical forms with a rusted old stapler. Centrifuged blood samples, drying on filter paper, were contaminated with fly droppings that contained cholesterol. The researchers politely accepted hospitality from locals who served octopus and dishes heavy with lard. But they also learned exquisite new dishes, like spinach lasagna, and simple foods like fresh pasta lightly seasoned with herbs and olive oil. They ate freshly harvested almonds and watched olives cold-pressed into oil. They were astounded at the lunch of a Finnish logger: huge chunks of meat surrounded by congealed fat, wrapped in dark bread soaked in more fat. Such meals, when analyzed, could contain as many as two thousand calories, mostly from fat.[12]

When Keys and his colleagues began, they knew only obvious things such as: the Japanese diet was low in fat, the U.S. diet was high in fat, and the Greek diet was high in olive oil. There were no data at the time about differences in fatty acids, about the fine points of distinction among monounsaturated fats, saturated fats, and polyunsaturated fats. Information about trans fatty acids, in foods like margarine, shortening, and most processed baked goods, was decades away.

They studied and reported on 12,770 men. After five years, 588 of those men had died, 158 of them from heart disease. In the United States, half (62 of 125) the deaths were due to heart disease; in Finland, 29 percent (11 of 38); in the Netherlands a third (16 of 50). For all the other groups combined, only one in eight deaths (12.5 percent) was due to coronary heart disease. The United States, where the men in the study were the fattest, had the second highest rate of heart disease, just behind Finland.[13] The Greeks, Yugoslavs, and Italians helped to show that not all fat is equal. Monounsaturated fat, abundant in olive oil, can be protective and healthy, while saturated fat in dairy and animal fats raises cholesterol levels and heart disease risk. The Finns demonstrated that even vigorous and strenuous daily exercise is not adequate protection against heart disease if it is accompanied by a high-saturated-fat diet. The Seven Countries Study is ongoing, and results keep coming in.

Jeremiah Stamler would follow Keys's lead and analyze diets in various countries. Indeed, the two men became friends and even neighbors at their summer homes in Pioppi, Italy. They shared the same cook, Dalia, who scolded them for eating canned foods when fresh foods tasted so much better. "My wife used to preserve fruit. Dalia would look at us like we were crazy. She'd say, 'You eat the fruits and vegetables of the season,'" says Stamler.[14] Dalia cooked seasonal vegetables, fish, and shellfish of the region in savory olive oil. Dessert was fresh fruit. In short, she changed their diets through delicious Mediterranean meals even as they were studying the beneficial effects of their new eating habits.

In the 1950s, Stamler was looking at cholesterol in chickens. He would drive from his research laboratory at Northwestern University north of Chicago to the HighLine Cockerel in Joliet, Illinois, and then drive back with a carload of chickens. He started with sixty-four birds, worked up to more than eight hundred, and spent Sundays cleaning their cages. He fed them fat, and produced high levels of blood cholesterol. He was still working with chickens when Felix Moore, a statistician from the National Heart Institute, came to see him. "We were feeding cholesterol to chickens, and producing atherosclerosis. Felix went away very skeptical," recalls Stamler. But not for long. Soon, data from Framingham volunteers fit with data from Stamler's chickens. "Two years later, I get a call from Felix saying, 'Can I come out and take another look at your chickens?' I asked why he was suddenly interested. And he said, 'Well, the Framingham data are showing a strong relationship of cholesterol to coronary risk.'"[15] It was becoming obvious that what was bad for chickens—and rabbits or monkeys in other labs—was also bad for human beings. And nutrition was finally becoming recognized as an important element in the prevention of heart disease.

The American palate, getting ever more demanding, has grown used to highly marbled meat from plump animals that get virtually no exercise. It is juicy and tender, and it bears little resemblance to the meat our farming and pioneering ancestors ate. On a trip to Cherbourg in the 1960s, Stamler remembers his first encounter with a different kind of meat. "We rented a car and began to drive until about one o'clock in the afternoon. Everyone was getting hungry. I came to a butcher shop, looked in the window, saw some meat, and didn't recognize any of it." It turned out ot be beef: pink, not bright red; lean, not marbled, and with only a thin skin of fat around the edges. It looked like no meat Stamler had ever seen in an American butcher shop. But it is probably the same kind of meat that Americans ate two or three generations ago, when cattle roamed and ran, when pigs moved outside in sties, before the advent of giant feedlots where animals are fed a fat-enhancing diet and their hooves touch only concrete. Cattle are bred and raised to provide us with meat that is tender and juicy because of its enhanced fat content. It could be that it's not only the quantity of meat that we eat, but the quality of meat that is killing us.

Stamler noticed other dietary differences in foreign populations. In Italy, he ate wild strawberries for the first time, and always had fresh fruits and vegetables in season. In Paris, he noted that the meal begins with appetizers, so the main course is small. Dessert is a sampling of fruit and cheese. In Japan, rice fills stomachs, and meat or fish is used sometimes as sparingly as a garnish, though salt consumption—like blood pressure and rates of stroke—is high.

Stamler and Keys, with similar research interests and a fervor for creating a new nutritional awareness, met in 1962 in Mexico City at the World Cardiology Congress. They became members of the conference's research committee. In 1966, when the congress met in New Delhi, they led the Council on Epidemiology and Prevention. "In a hot convention room in New Delhi, with no air-conditioning, we sat around and asked, 'What are we going to do?'" says Stamler. "We decided to publish a newsletter to get more information out on epidemiology and prevention. We talked about organizing international research. We knew we needed troops. We created a ten-day international seminar on cardiovascular disease prevention, the first one in Yugoslavia. We had it there because Keys was there for the Seven Countries Study." They mapped out a strategy for letting the medical world know that what goes into the stomach affects the heart.[16]

Stamler went on to lead one of several long-term epidemiology studies of lifestyle and health, the People's Gas Company Study in Chicago, begun in 1958 and so named because volunteers worked for the utility company. "We took it up and proposed a screening project in industry to screen for risk factors. It had a very traumatic start. We began in the first company recruiting men only. Within twenty-four hours, there was a reaction from women screaming bloody murder: Why are you discriminating against women? But the thinking of that era still was that it was overwhelmingly a disease of men," he says. "Based on that reaction from women the first day, we opened it up for men and women. Thank God for those [complaining] women because now we have data on women."

Oglesby Paul was starting up a similar look at the lifestyles of workers at the Western Electric plant in Cicero, Illinois, just west of Chicago. Eventually, results from five epidemiology studies were pooled, an effort organized by James Watt, director of the National Heart Institute. He paired enthusiastic young researchers with cardiology chiefs. They met at Arden House on the Harriman estate in upstate New York, and after the usual scientific nervousness about sharing forms and pooling data, they agreed to collaborate.[17] The pooled data came from the two Chicago studies as well as the Framingham Heart Study, and civil servants from Albany, N.Y., and Los Angeles. As years went by, the results confirmed animal findings about dietary fat, cholesterol, and heart disease. If the Framingham Heart Study is the landmark for defining risk, these studies dotted the *i*'s and crossed the *t*'s about diet, smoking, and exercise.

Smaller dietary investigations were cropping up around the country. One group studied Zen Buddhists in Boston; another observed Seventh-Day Adventists in Loma Linda, California.[18] All were looking for discernible effects of low-fat, vegetarian, nonsmoking, nondrinking lifestyles to compare with the typical American overindulgence.

Meanwhile, the Framingham participants were answering dietary questions. They, like most Americans, were eating stews, creamed tuna, meat loaf, corned beef and cabbage, mashed potatoes with lots of butter. They were consuming breaded and fried veal cutlets, deep-fried cauliflower, and French fries. "I remember my mother would pan-fry potatoes and baste them with Crisco. She had a deep-fry cooker that would sit on the counter like you might have a toaster now," says Karen LaChance. Meat was on the table five or more nights a week. Lamb chops. Liver. Cheesecake for dessert. Real butter. Whole milk. "There was no concept about the harm. This was considered good food," says Larry Shapiro, another Heart Study participant. "Greens and ham hocks. Fried chicken. Country ham," recalls volunteer Deb Fuller. All of these various feasts, people were certain, represented good, healthy nutrition.

With countless ethnic variations, they were all eating a high-saturated-fat, high-cholesterol diet, blissfully certain it was good for them, just as researchers were beginning to pinpoint that kind of diet as a factor in increasing the risk for heart disease. A 1952 American Heart Association booklet offered a tepid attempt at balance. It said, "Most Americans prefer animal protein such as meat, fish, milk and eggs. But the vegetable protein of beans, peas, wheat, corn and oats is quite satisfactory." The booklet made no differentiation or recommendation on fats: "Everybody needs a little fat. Enough may usually be had from milk, eggs, meat, fish and fowl."[19]

By 1961, the same year as Kannel and Dawber's paper "Factors of Risk in the Development of Coronary Heart Disease" from Framingham, the American Heart Association published the first brochure on prevention that specifically mentioned dietary fat reduction as a way to reduce risk. The recommendations still weren't clear, but they were getting closer to the mark:

> The question many of us would like to have answered is: will cutting down or changing the kinds of fats we eat lessen the risk of a heart attack or stroke?
>
> There is no final proof that changing the fat content of our diet will prevent heart attacks or strokes caused by atherosclerosis. However, based on present knowledge, some recommendations have been made for those who may be prone to these conditions. Keep in mind that important changes in diet should be made only with your doctor's advice.
>
> What are the dietary changes recommended?
>
> 1. Eat less fat.
>
> 2. Substitute a substantial amount of liquid vegetable oils for the solid animal fats, such as butter and the fat in meat. . . .
>
> Although normal weight is no guarantee against heart trouble, doctors are firm on this point—overweight is a health hazard, whether you have heart disease or not.[20]

For the next decade and a half, the advice was temperate, with no sense of national urgency. The Senate Select Committee on Nutrition and Human Needs changed all that. In studying hunger in America, the committee also found the other end of the spectrum: overeating and overweight. Its report on "Dietary Goals in the United States" was released in 1977. For the first time the government urged a *decrease* in consumption of American-grown foods, including meat, eggs, and dairy products, which until then were supported and promoted by the Department of Agriculture. It was the first comprehensive statement by any branch of the government on risk factors in the American diet.

Politics had entered the arena and legislators from farm states, like Robert Dole of Kansas and George McGovern of South Dakota, had to deal with egg, dairy, and cattle producers who feared that their foods would be labeled "unhealthy." Senator Bob Dole once told a gathering in his home state that farmers get heart attacks, too, and stood firm on the 1977 dietary recommendations.[21]

And Senator George McGovern, then chair of the Senate Select Committee on Nutrition, reported what researchers in Framingham and elsewhere had been observing for years. "The simple fact is that our diets have changed radically within the last 50 years, with great and often very harmful effects on our health," McGovern said on January 14, 1977. "These dietary changes represent as great a threat to public health as smoking. Too much fat, too much sugar or salt, can be and are linked directly to heart disease, cancer, obesity and stroke, among other killer diseases. In all, six of the ten leading causes of death in the United States have been linked to our diet."[22]

Senator Charles Percy of Illinois, ranking minority member of the committee, added: "Without government and industry commitment to good nutrition, the American people will continue to eat themselves to poor health. . . . Our national health depends on how well and how quickly government and industry respond."[23]

Publication of the report pushed the government to continue with recommendations known as the U.S. Dietary Guidelines, and for the first time Americans were told officially to eat more fruits, vegetables, whole grains, poultry, and fish; and to cut down on foods high in fat, as well as to substitute nonfat milk for whole milk. From that point on, and despite the growing obesity epidemic, few could argue that they didn't know better.

A debate persisted over whether obesity per se was a risk factor for heart disease. Throughout the 1970s, it was commonly believed that overweight itself was not a culprit. Rather, obesity was thought to be a benign consequence of sedentary lifestyle and nutrition habits that caused elevated cholesterol levels, both of which contributed to heart disease. But Kannel and Castelli figured otherwise. They were noticing that among the Framingham participants, those who were overweight had higher blood pressure levels, and the more overweight they were, the more elevated the pressure.

In 1983, a Framingham publication was titled "Obesity as an Independent Risk Factor for Cardiovascular Disease: A 26-Year Follow-Up of Participants in the Framingham Heart Study."[24] It marked the beginning of the end of thinking that obesity by itself was harmless.

By 1984, the National Institutes of Health was advising a restriction on fat consumption, and the U.S. surgeon general followed with a report in 1988 on the dangers of dietary fat. By 2000, the government, in a document called "Report on Nutrition and Health," called fat the single most unwholesome component of the national diet.

In large part, Americans have listened, heard, and understood, with years and years of repetition and clarification on how to eat. But recommendations remain confusing to many, and the science of nutrition continues to be controversial. A recent ripple was a July 7, 2002, cover story in the *New York Times Sunday Magazine* by Gary Taubes. Few who saw the cover photograph of a T-bone steak topped with a generous pat of melting butter will forget it. It was an appetizing representation of the postwar meat-and-potato years, a plateful many Americans would love to return to without guilt. In his article Taubes appeared to give them the chance by suggesting that public health support for low-fat diets was wrong. Instead, he offered the Atkins diet—a low-carbohydrate, high-fat diet allowing for large portions of fatty meats and only small amounts of vegetables, and almost no fruits and grains—as a healthy alternative for weight loss.

I cringed, as did Castelli, Kannel, Dawber, and countless scientists who had spent their careers understanding and proving the dangers to the heart of a high-saturated-fat diet. Within weeks, publications including the *Washington Post* examined Taubes's article, citing evidence that contradicted or refuted key elements. And recently, even officials at Atkins Nutritionals, which sells and promotes Atkins products, appeared to back away from advising unlimited fat by recommending no more than 20 percent of daily calories from saturated fat, which is still far more than the current recommended daily limit of the American Heart Association and other organizations.

At last, however, the controversy is beginning to be put to the test in clinical trials. A randomized study presented at the American Heart Association Scientific Sessions in November 2003 compared four popular diets: Atkins (low carbohydrates), Zone (moderate carbohydrates), Ornish (very low-fat vegetarian), and Weight Watchers (moderate fat). Volunteers were assigned to one of the diets, then left on their own to follow the plan. They ran into the chronic dieter's dilemma: half the volunteers on the Ornish and Atkins diets dropped out after a year, as did 35 percent of those on the Weight Watchers and Zone diets.

Those who stuck with any program lost weight. And, using the Framingham risk models, they lowered their predicted heart disease risk scores by differing amounts: Weight Watchers, 14.7 percent; Atkins, 12.3 percent; Zone, 10.5 percent; and Ornish, 6.6 percent. All but the Ornish diet significantly increased levels of protective HDL cholesterol.[25] Matching diet to lifestyle might help people stick with weight loss efforts. But the long-term effects of a high-protein, high-saturated-fat, low-carbohydrate diet remain unknown.

The road to identifying dietary fats, and sorting the good from the bad, has indeed been bumpy and long. But today it's clear that what makes sense is lowering total fat intake to about 30 percent of calories, and concentrating those fats into monounsaturates found in olive and canola oil, and in moderate helpings of treats such as nuts, olives, and avocados, and frequent meals of fish. In research involving 22,043 adults in Greece over forty-four months, questionnaires aimed to get at adherence to the Mediterranean diet (high in fruits, vegetables, legumes, olive oil, and fish; low in saturated fats, meat, and poultry, and moderate in dairy products). Those who followed the Mediterranean eating plan had fewer deaths from heart disease.[26]

The average American still falls far short of a healthful lifestyle. Obesity is the new American epidemic. Next to smoking, obesity is the second leading cause of preventable death in the United States, resulting in 300,000 lives lost each year.[27] More than 64 percent of Americans over the age of twenty are overweight, and 30 percent are obese.[28] That means a body mass index of more than 30, roughly corresponding to a waistline measuring more than forty inches for a man, thirty-five inches for a woman.

Lifestyle studies have shown that to be at the lowest risk for heart disease, a person must have blood pressure less than 120/80, have total cholesterol under 200, not smoke, and not have diabetes or a family history of heart disease. Only 10 percent of adult males and 20 percent of adult females in the United States are at that low level of risk. Everyone else is at moderate or higher risk of heart disease.

Henry Blackburn, soon after analyzing results of the Seven Countries Study in Crete in 1970, described the low-coronary-risk male. In his prose, someone who lives a heart-healthy life sounds like a person returning to Eden.

> He is a shepherd or small farmer, a beekeeper or fisherman, or a tender of olives or vines.
>
> He walks to work daily and labors in the soft light of his Greek isle, midst the droning of crickets and the bray of distant donkeys, in the peace of his land.
>
> At the end of his morning's work, he rests and socializes with cohorts at the local café under a grape trellis, celebrating the day with a cool glass of lemonade. . . .
>
> He continues the siesta with a meal and a nap at home, and returns refreshed to complete the day's work.
>
> His midday, main meal is of eggplant, with large livery mushrooms, crisp vegetables and country bread dipped in the nectar that is gold Cretan olive oil.
>
> Once a week there is a bit of lamb, naturally spiced from grazing in thyme-filled pastures.
>
> Once a week there is chicken.
>
> Twice a week there is fish fresh from the sea.[29]

Blackburn's low-risk islander also eats legumes, salads, fruits, dates, and nuts, and caps his dinner with a glass of fragrant local wine.

Unfortunately, we can't all live on a Greek island. And Americans are much more ready to believe in the next diet craze than to sort out the differences between trans-fatty acids and monounsaturated fats. They are more willing to hope for a magic pill than to walk to the super-market or take the stairway instead of the elevator. In many other parts of the world, those who eat differently from Americans maintain healthier lives. Foreigners visiting this country are amazed that food is everywhere—in airports, railroad stations, stands on city streets, and food courts in suburban malls. Children open the refrigerator door all day long. In a 1997 book *Fat History: Bodies and Beauty in the Modern West,* Peter Stearns writes that the number of diet books in print quadrupled from 1959 to 1983. An Amazon.com search for diet books comes up with nearly 12,000 titles. It's ironic that this coincides with an increase in the body mass index—the most accurate way of measuring overweight and obesity—of the U.S. population.

The epidemic of obesity is confounding all that has been learned about controlling the epidemic of heart disease. It is proving to be the most stubborn obstacle to healthy hearts, and the statistics keep going in the wrong direction. As weight increases, so do blood pressure and cholesterol levels. So does diabetes. And combined risk factors create far greater danger than any one factor alone. Add smoking, and peril multiplies.

The accumulation of poundage escalated throughout the last forty years of the twentieth century. From 1960 to 2000, overweight and obesity rates in adults climbed from 45 percent to 64 percent. More than 15 percent of children aged six to nineteen were overweight in the year 2000, and an additional 15 percent were considered at risk for overweight.[30] We haven't even begun to see the repercussions of rising levels of obesity in children. No one is immune. Those with college educations and adolescents packed on the pounds along with the more traditional overweight groups of middle-aged adults and people low on the socioeconomic scale. The lifestyle risk factors are entwined and negate much of the success of fifty years of discoveries on preventing cardiovascular disease. The death rate from heart attacks and strokes has gone down, but could be reduced much further by following the wisdom gleaned from research.

Millions are still oblivious to the silent risks. But, thanks in large part to Framingham, they know enough to feel guilty. "I eat the wrong things, but when no one is looking," says Evelyn Langley, now in her late eighties.

The science behind lifestyle risks has been rife with controversy, a confusing swirl of contradictory research and vested interests. At the turn of the twentieth century, the federal government, in good faith, set a standard of 4 percent fat in milk in order to qualify for the label "whole milk." The thinking then was that whole milk was the healthiest, and farmers were skimming the milk, taking out the fat to sell as cream. No one knew then that skim milk was a healthful by-product. Researchers, also in good faith, lectured on the heart-healthy benefits of ice cream, believing at the time that its calcium content lowered blood pressure.

Following early results on cholesterol, the National Commission on Egg Nutrition ran advertisements saying there was no proof that eggs caused harm—until a false advertising suit in 1977 by the Federal Trade Commission stopped the campaign.[31] The link between egg consumption and cholesterol levels remains controversial even today. Not all fats are equal, and not all are equally bad. For every move in public health education about meat, dairy products, eggs, fat, and salt, there has been a powerful voice from the beef, dairy, egg, and salt industries attempting to discredit the studies.

But the science continues to add up. From Framingham, from the Seven Countries Study, from workers at Western Electric and People's Gas Company, from Loma Linda, from the Physicians Health Study, from Helsinki and Italy and India and Holland and Japan, studies point the way to a healthy diet geared to the prevention of heart disease. A diet that includes an abundance of fish, rice, fruits, and vegetables, a modest amount of monounsaturated olive or canola oil, and perhaps a moderate splash of alcohol is likely to help protect the heart. Soy and oat bran can help lower cholesterol, as can new classes of margarinelike spreads containing plant sterols.

Americans have shown that they can reduce their consumption of fat. A full-blown junk food binge of cheeseburger, fries, and a milk shake feels like something akin to sin. In 1965 we were getting 45 percent of our calories from fat. Now it's down to 32 percent—just two percentage points above national recommendations,[32] but still not what some would argue is optimal: 25 to 26 percent. But are we really eating less fat? When fat consumption is measured in daily grams rather than as a percentage of total calories, Americans have been adding grams since about 1990. The percentage of dietary fat has gone down in part because a higher number of calories eaten will reduce the calculated percentage of calories from fat even when there is no actual decrease in total fat consumption.[33]

Americans are fat, and getting fatter, even though we know better. More than seven in ten adults say it's important to them to maintain a healthy weight.[34] Yet they continue to pile on the calories. For most, the way to lose weight is by following a simply stated strategy dictated by the laws of thermodynamics: Consume fewer calories than you expend. It's enormously difficult to execute that strategy. The Department of Agriculture surveyed Americans' eating habits, and found that 70 percent of them fell into the "needs improvement" category, while only 12 percent had a "good" diet and 18 percent had a "poor" balance of nutrients.[35]

Up through the 1970s, a commonly held belief even among scientists was that obesity itself had nothing to do with heart disease. Obese people were more likely to have high blood pressure, elevated levels of total cholesterol, low levels of HDL cholesterol, and high rates of diabetes, but the "independent" role of obesity was considered by some to be inconsequential. That illusion came to an end with a 1983 Framingham report linking obesity to heart disease risk.[36]

All of this information received attention, though it's not what anyone wants to hear. A memorable newspaper photograph in 2000 of two laboratory mice from a Johns Hopkins lab—one tubby, the other pencil thin—came close to realizing the dream of a magic diet pill.[37] The image of the fat rodent beside his svelte comrade really had Americans abuzz with hope. The obese mouse had a flawed gene and was unable to produce leptin, a protein seen as key to appetite control. The skinny mouse was fed a substance called C75 that shut down its appetite and led to a loss of 30 percent of total body fat. The research has not yet produced the miracle treatment many undoubtedly long for. It *would* be so much easier than a healthy diet and exercise—the prudent approach that eludes so many Americans. John DeCollibus, a Framingham Heart Study participant, was sixty-five years old, five feet nine inches tall, and 185 pounds in 1982 when he told *Science 82* magazine: "I feel like a dummy walking up and down the street, and jogging is overkill. Man was given the ability to run to avoid dangers, not to dash around the neighborhood." More than twenty years later, he still speaks for the masses.

Exercise is nutrition's comrade in health. Studies going back to 1953 showed that exercise is protective against cardiovascular disease. London bus drivers were at a greater risk than conductors, who spent their days walking the length of buses and running up and down the stairs of the city's double-deckers. Postal workers who delivered mail had less heart disease than those who worked behind the counter. Longshoremen in San Francisco with moderate or high levels of leisure-time physical activity had lower death rates than those who were sedentary.[38]

Paul Dudley White was among a handful of physicians who believed in the value of exercise long before any proof came in. He set an example by riding a bicycle, and spoke about its likely benefits. "Walking is probably the best exercise because it is easy for anyone to accomplish," he said, according to his biographer Oglesby Paul. Before science backed him up, he was urging a choice of stairs, not elevators. And, following Eisenhower's heart attack, when he was to speak at the National Press Club, White took his own advice, bypassing the waiting elevator and walking to the thirteenth floor of the building in Washington.[39]

But through the 1950s and 1960s, exercise was controversial. A lot of cardiologists thought the association between exercise and a reduced risk of heart disease was just coincidental, and there were others who believed strenuous exercise was downright dangerous and unhealthy. It wasn't until 1978 that the College Alumni Health Study found that among 50,000 Harvard College and University of Pennsylvania alumni there was an inverse relationship between physical activity level and risk for heart disease.[40] It clearly showed that those who were physically active lived longer. Even more encouraging, it found that those who changed from a sedentary to an active life also benefited.

College Alumni Health Study leader Ralph Paffenbarger, aged eighty in 2002, is a fine example that it's never too late to start exercising. He has run more than 150 races of marathon length or longer. He's run from Hopkinton to Boston, from London to Brighton, the Two Oceans Marathon in South Africa, and from Squaw Valley to Auburn, California. Once, he ran the London Marathon, showered at Heathrow Airport, flew to Boston, and was picked up and driven to Hopkinton, where he ran the marathon back home to Boston. But he began running only when he was forty-five years old. In 1960 he was a researcher at the Framingham Heart Study when data began to suggest that those who were more active had less heart disease. Framingham results, based on a fairly sedentary group of participants, were inconclusive. Paffenbarger didn't begin exercising until results from the College Alumni Health Study started coming in.

"The day after Thanksgiving in 1967, I decided that it was time for me to begin a jogging program," he says. His gear was not what anyone would call sophisticated. "I put on my Army boots and headed out the front door early so my neighbors wouldn't see me. I ran to the end of the block and was exhausted." But he persisted, running every day, rain, shine, or snow, right on up to the New England holiday of Patriots' Day. "The night before the Boston Marathon, the nineteenth of April in 1968, I announced to the family I needed somebody to drive me out to Hopkinton (the race's starting point) so I could run the marathon. We didn't know anything about running in those days. I switched to deck shoes, and I ran it in five hours and five minutes."

The College Alumni Health Study has shown that you don't have to run a marathon, or even jog at all. The optimal amount of exercise is still uncertain. How long, how often, and how hard do you have to work out for maximum benefit? Paffenbarger is unsure:

> This is tough. We've worked on this forever. If there is something beyond the optimal level of activity, we don't know what that is. There must be some level that's too much. My friends say that I got too involved. I've run a hundred miles five times. There must be an optimal level for people of given ages, with given circumstances, habits of diet, weight, mood, and psychological behavior. We can't come up with an ideal amount. But quantity and intensity are both important. We go with the idea that's been prescribed that people should spend thirty minutes a day doing things that are moderately vigorous most days, and preferably all days, of the week.[41]

Thirty minutes a day of something as simple and available as walking will help prevent heart disease. And research shows that if the pace of exercise feels right, it probably is. In a study of eighty-four obese adults who were told to walk on a treadmill at a pace that feels brisk but comfortable, the self-paced walkers reached recommended levels of intensity, or a heart rate of 55 percent of its maximum. They didn't have to jog to get there. The average pace was 3.2 miles per hour.[42] It's important to remember that with exercise, something is better than nothing. Ten minutes here or ten minutes there can make a difference.

Paffenbarger is still involved in exercise studies, though he has had two heart attacks and has an implanted automatic cardiac defibrillator. Now he walks three to four miles a day. He says he will always miss running: "I had to stop very suddenly. I have a recurring dream of running."

In 1992, the American Heart Association published a position statement on exercise: "There is a relation between physical inactivity and cardiovascular mortality, and inactivity is a risk factor for the development of coronary artery disease." In 1995, the Centers for Disease Control joined forces with the American College of Sports Medicine to recommend, as Paffenbarger did years before, thirty minutes or more of moderate physical activity on most, preferably all, days of the week. The surgeon general echoed that recommendation in 1996.

Yet millions of Americans shun exercise. A quarter of all trips in this country of less than a mile are taken by car. Evelyn Langley is not unusual for her generation in recalling treks to school in all weather. Today, two-thirds of children who live within a mile of their school get a ride. Despite the popularity of health clubs and the increase in the number of marathons and marathoners, 25 percent of Americans are sedentary, and more than 60 percent don't get the minimum recommended amount of exercise.[43]

FOURTEEN

Deadly Addiction

At the turn of the twentieth century, fewer than one in one hundred Americans smoked regularly.[1] But the generations coming of age thereafter became hooked on nicotine in blissful ignorance. President Franklin Roosevelt looked dapper with the familiar cigarette holder in his mouth. Thousands of veterans came home from World War II with newly acquired nicotine addictions, and by the time the Framingham Heart Study began, more than half of adult men smoked. William Kannel was bumming cigarettes from Pat McNamara as they planned and executed the analyses that would persuade them to stop smoking. Henry Blackburn, traveling in Yugoslavia and Greece for the Seven Countries Study, was sampling local hand-rolled cigarettes, and longing for the smoother nicotine hit of American brands.

The tobacco industry's marketing efforts of the day are breathtaking in their show of arrogance and deceit. Magazine ads for Camels— ahead of their time in the occasional use of a female doctor—were implying the safety of their product and claiming that more doctors smoked Camels than any other cigarette. L&M had an advertising campaign that called its cigarettes "Just what the doctor ordered." Viceroy claimed to give "double barreled health protection." Today a physician, or a U.S. president, would be loath to be seen smoking. Public policy has sent smokers outdoors or to the privacy of their homes. Using nicotine patches or gum, hypnotism, or sheer willpower, and pressured by smoke-free offices, restaurants, and even bars, smokers are quitting and adult nonsmokers know they shouldn't start.

But in 1962, when Framingham scientists published findings showing that cigarette smoking increased the risk of heart disease, it was big news.[2] Lung cancer had been recognized as a consequence of smoking since 1952, but it took fourteen years of examinations of Framingham participants to prove the even more widespread cardiovascular risk of tobacco.

Cigarette smoking, a deadly risk factor, still has a stranglehold on tens of millions. Despite the fact that it is a controllable addiction, and that hardly a human being alive can claim ignorance about tobacco's harmful effects, nearly one-quarter of adults continue to smoke.[3] Worse yet, children in their teens and even younger gladly suffer through the initial coughing and the burning pain of smoking in the name of what they believe to be sophistication and defiance of authority.

As countless books have documented, including Richard Kluger's Pulitzer Prize–winning *Ashes to Ashes* and Philip J. Hilts's *Smokescreen,* the industry effectively confused and contradicted scientific findings about the harm caused by tobacco products. The movie *The Insider* depicted the tobacco industry as being on a par with the Mafia for kindness and empathy.

But that information came out slowly.

The answers found in Framingham helped set public policy on warning labels required on cigarette packages, and were a significant factor leading to the first surgeon general's report on smoking.[4] Kannel served on the Surgeon General's Advisory Committee on Smoking and Health and remembers meetings before release of the famous report.[5] "We were in a basement in Bethesda, at the National Library of Medicine. The surgeon general and all the government players were around the table," he recalls. In chairs ringing the room were representatives of the tobacco industry holding clipboards and studies done by their own scientists. "They weren't *at* the table, but they were sure *near* the table."

In public statements, the Council for Tobacco Research, the industry's arm, claimed to be dedicated to dealing with the "health scare" presented by smoking. Its method, broadly documented, was to fight scientific findings with scientific-sounding attacks. Some of the latter were cheap shots. One from the Cold War era, for example, implied that cigarette critics had hidden pyrophobia and a repressed fear of the atom bomb.

But for the most part, the industry used the language of science and spread a smokescreen of respectability. In response to Ernest Wynder's report of tumors induced in mice by painting cigarette tar on their backs, the Tobacco Industry Research Committee, later known as the Council for Tobacco Research, was formed in 1954. It issued "A Frank Statement to Cigarette Smokers" and placed it in ads in 448 newspapers reaching a circulation of 43.2 million readers.

Recent reports on experiments with mice have given wide publicity to a theory that cigarette smoking is in some way linked with lung cancer in human beings.

Although conducted by doctors of professional standing, these experiments are not regarded as conclusive in the field of cancer research. However, we do not believe results that are inconclusive, should be disregarded or lightly dismissed. At the same time, we feel it is in the public interest to call attention to the fact that eminent doctors and research scientists have publicly questioned the claimed significance of these experiments.

Distinguished authorities point out:

That medical research of recent years indicates many possible causes of lung cancer.

That there is no agreement among the authorities regarding what the cause is.

That there is no proof that cigarette smoking is one of the causes.

That statistics purporting to link cigarette smoking with the disease could apply with equal force to any one of many other aspects of modern life. Indeed the validity of the statistics themselves is questioned by numerous scientists.

We accept an interest in people's health as a basic responsibility, paramount to every other consideration in our business.

We believe the products we make are not injurious to health.

We always have and always will cooperate closely with those whose task it is to safeguard the public health.

For more than 300 years tobacco has given solace, relaxation, and enjoyment to mankind. At one time or another during those years critics have held it responsible for practically every disease of the human body. One by one these charges have been abandoned for lack of evidence.

Regardless of the record of the past, the fact that cigarette smoking today should even be suspected as a cause of a serious disease is a matter of deep concern to us.[6]

That was just the opening round of a battle between independent science and tobacco industry marketing efforts.

Dwight D. Eisenhower's heart attack on September 27, 1955, made history. In stark contrast to the medical cover-up surrounding Roosevelt, Americans heard the truth about the health of their president at the time of his illness. Two years later, Eisenhower would have an apparent stroke. From then on, he would experience heart problems, eventually dying of cardiac disease in 1969. During the war years and his presidency, he smoked like the proverbial chimney—sixty cigarettes a day or more.[7] Today, a three-pack-a-day habit is known to be self-destructive, a kind of slow suicide. But in those days, the full danger of tobacco was not widely understood.

In 1957, researchers showed that premature births were twice as likely among women who smoked as among nonsmoking women.[8] At the time, the tobacco industry was still providing free cigarettes at annual medical and public health meetings.

In a 1963 Framingham report, Kannel wrote that confusing findings with moral judgments about tobacco set research back. Some people were skeptical of the harmful effects of smoking "because of the obvious fallacy of the earlier 'old wives' tales' about the effect of tobacco smoking on stunting growth, etc." Another obstacle to putting smoking on the list of risk factors was that smokers had a higher rate of death from a large number of illnesses, not just heart disease. Exceptional death rates from lung diseases and cancer in fact partly masked the association of smoking with heart disease. Smokers simply didn't live long enough to develop heart attacks in large enough numbers to draw definitive conclusions early on.

In 1963, Kannel wrote:

Whether adults capable of making their own decisions as to what hazards they may wish to risk in the search for personal pleasure require any further approach than being advised of the facts concerning the danger of cigarette smoking is a matter that involves more than medical opinion. There is considerable hazard in parachute jumping or sports car racing but many enjoy these activities and are free to engage in them in spite of well-known risks.

The fact that it is in childhood that the cigarette habit is acquired and that the decision to take up smoking is made by children who are "sold a bill of goods" by advertisers and by example of adults not acquainted with the deleterious consequences of cigarette smoking should cause physicians to consider their responsibility in helping to combat and counteract the propaganda aimed at converting large numbers of children into life-long consumers of cigarettes.[9]

A meticulous scientist, Kannel uncharacteristically went out on a limb with his editorial opinion. But he was right. The tobacco industry hooks 89 percent of its lifetime customers before they turn nineteen.[10]

The 1964 Surgeon General's Report on Smoking and Health was the first official recognition that cigarette smoking can lead to cancer. The Advisory Committee concluded that it was a cause of lung and laryngeal cancer in men, a probable cause of lung cancer in women (who had begun smoking later in the century than men and were not yet getting lung cancer to the same extent), and the leading cause of chronic bronchitis. The report was released on a Saturday to forestall a reaction on Wall Street. The press received copies of the 387-page report in a State Department auditorium. They were locked in with no telephones, and were given ninety minutes to read and digest the report and ask questions.[11] When reporters were released to file stories, the news made headlines.

The Tobacco Institute responded with its own salvo. It reported that efforts to induce human-type lung cancer in animals through the inhalation of tobacco smoke had uniformly failed.

It wasn't until 1967 that an updated surgeon general's report included heart disease among the illnesses caused by cigarettes.[12] The Framingham findings, which contributed to that conclusion, were the result of years of analyzing answers from Study participants to dozens of questions on cigarettes: How long have you smoked? How many cigarettes do you smoke? Filtered? Unfiltered? Do you smoke the whole cigarette? Three-quarters? Half? One-quarter? Regular or low-tar brand? How old were you when you started? If you quit, when did you quit?

By the time the Heart Study found that smoking was a risk factor for heart disease, Carl Seltzer of the Harvard School of Public Health was emerging as the chief critic of Framingham results. A heavy smoker himself, he built a career on finding flaws in the study's research methods. "I don't know why he did it," Kannel says. "I had a conversation with him once. I said, 'Look, is this really how you want to be remembered? As the guy who promoted the value of cigarettes?'" Maybe his own heavy habit helped Seltzer ignore the dangers of tobacco.[13] More likely, it was the money the Tobacco Research Council could provide. One letter on file with tobacco.org is a recommendation that Seltzer's contract be renewed in the amount of $70,000 for fiscal year 1989. As part of the recommendation, Donald Hoel, of the law offices of Shook, Hardy & Bacon in Kansas City, Missouri, cited Seltzer's work analyzing the Framingham data and focusing on "methodological issues which may pose problems in the interpretation of Framingham data." In other words, he was highly paid to put up a smokescreen.[14]

Castelli remembers that Seltzer "would take the Framingham data and dig out a subgroup, like plain angina, meaning he tossed out 80 percent of the cardiovascular events. He was always nursing the data, pulling out one little thing, to show that smoking was okay." But cigarettes are such a horrendous risk factor that ultimately no amount of data distortion could make smoking look benign. "I used to bump into him once in a while," Castelli recalls. "I'd say, 'Hey, why don't you instead pull out the sudden death stuff?'" In the Framingham data, cigarette smoking was highly correlated with sudden death from heart attack. "Cigarettes were so bad. The first big finding was in young men who dropped dead suddenly. Young men were dropping dead all over Framingham, and the one big thing they had in common was smoking cigarettes."[15]

Seltzer wasn't alone. People calling themselves "independent" researchers would write to the National Institutes of Health asking for unpublished Framingham data. One such letter prompted an internal memo on November 18, 1969, from Tavia Gordon of the National Heart Institute's Biometry Research Section. "Mr. [name deleted], so far as I can tell, is an entrepreneur who has no data-collection system of his own and whose specialty is computers. He is apparently funded in part by the Tobacco Research Institute." Gordon went on to say that others inside the NHI considered the letter writer an "incompetent analyst."[16]

For years the industry, with the assistance of hired guns like Seltzer and entrepreneurs who would file Freedom of Information Act requests with the National Institutes of Health, fought finding after finding about the harmful effects of tobacco. They ran commercials using cartoons and advertisements claiming physician endorsements. With characters like Joe Camel and the Marlboro Man, no doubt millions of children were hooked on images they saw as sophisticated or macho. The industry touted the relaxing effects of smoking, the smooth draw of tobacco, and finally the "safety" of filtered cigarettes. If cigarettes were bad—and the industry was not saying they were bad—filters would reduce the risk.

William Castelli was intent on doing a study analyzing the risk of filtered cigarettes. Others tried to talk him out of it. Robert Garrison, then at the National Heart, Lung, and Blood Institute, believed, along with many others, that smoking filtered cigarettes might reduce the risk of harm, and he didn't want a Framingham finding to give more ammunition to the tobacco industry.[17] But Castelli didn't buy conventional wisdom, and went ahead.

"We didn't know if filters made smoking safer or not," says Castelli. "But we didn't give a darn what anyone else was going to say or do about any of our findings. We wanted to find the answer, and we just didn't know. We're data people. So we started monitoring carbon monoxide [an indicator of smoking] in expired air in everybody [Framingham volunteers]. It looked like the people on these filtered cigarettes would actually inhale deeper because it wasn't that irritating. They actually ended up with higher carbon monoxide readings." To everyone's surprise, Castelli found that filters on cigarettes did *not* reduce their harm.[18] In 1981, a Framingham publication terminated the industry's ability to promote filtered cigarettes as "safer."

In 1988, Framingham also documented a link between smoking and stroke.[19] When the Heart Study began, some 44 percent of Americans smoked—54 percent of men and 33 percent of women. Today, about 23 percent of Americans smoke, with women at nearly the same rate as men.[20] But the decline is beginning to level off among adults. A disturbing trend has been slowed but not stemmed in adolescents. Smoking among young adults aged eighteen to twenty-four rose steadily from 26 percent in 1993 to nearly 37 percent in 1997. But by 2001, rates among that group had dropped to 29 percent.[21] The fact that nearly one in three young people takes up the habit does not bode well for their future health.

The science on healthy lifestyle keeps getting clearer all the time. Eat a low-fat, nutritious diet. Exercise. Don't smoke. Monitor your blood pressure and cholesterol level. The American Heart Association estimates that if all cardiovascular diseases were eliminated, life expectancy in the United States would rise by almost seven years and the nation's health care expenditures would drop by more than $300 billion.[22] Imagine the dent in disease rates if all of us simply did what we know we should do. But unfortunately that's not going to happen anytime soon.

FIFTEEN

Spreading the Word

To hear William Castelli describe the innards of a hot dog is enough to make you swear them off forever. "You know what they put in those things, don't you? Start with the organs we can't mention in public and work your way up to the eyeballs," he says. Castelli has used variations of his description of hot dog ingredients for years in a class for Harvard Medical School students held at supermarkets. One medical student who took the class met Castelli years later and swore he hadn't looked at a hot dog since.

When it comes to spreading the word on preventing heart disease, Castelli is in character twenty-four hours a day. In the press, his quotations are often identical, even when found in different publications and given to a variety of reporters. His carefully chosen words come from an intrinsic awareness of how to connect with both the medical community and the lay public in delivering his message of heart health.

I have heard his preaching delivered casually by the coffeepot at the Framingham Heart Study headquarters; I have heard it in community hospital basements before an audience of no more than ten; I have heard it at international medical conferences where some of the world's leading cardiology experts have challenged his message. So dedicated is Castelli to being the Heart Study's emissary that he can sound, to those of us who have heard him often, repetitive and predictable.

And yet I have also seen the reaction on the faces in his diverse audiences. He connects. He has mastered the sound bite and, through the almost obsessive transmission of his message, he reaches the people who can benefit from knowledge of risk factors and from pointers on how to minimize the ones they can control. Some have characterized his presentation style as theatrical or evangelical. Perhaps so, but one thing is certain. He knows how to sell the message of heart disease prevention better than anyone else I've ever encountered. Castelli has always wanted to get the word out, to publicize the news of risk factors, to attract the public's attention and impress on them: *Watch what you eat. Exercise more. Don't smoke.* He remains a fervent public health messenger.

When I think of James Herrick, the researcher who discovered in 1912 that some patients could survive a heart attack, I see similarities to Castelli. Both were more than ready for medical science to start disseminating the news. Neither could wait to take medical findings on the road to begin trying to persuade physicians and individuals to change their habits. But Herrick was frustrated by having found a medical clue and no solution. He couldn't get the attention of his colleagues, much less the general public. Castelli has learned that when it comes to findings on how to prevent heart disease, his medical peers and the general public speak the same language.

He has also learned, along with the rest of the nutrition community, that the public needs continual reminding and updating. For a long time, fat was seen as the major culprit in the American diet. And from that assumption grew a steadily increasing variety of fat-free and low-fat products. Some such alternatives, like skim milk, are wonderful additions to mainstream supermarkets. But others, like fat-free cookies, cakes, and ice cream, have given Americans the false reassurance that they can eat as much as they want. Calorie intake has gone up just as fat content has declined in the average diet. And the obesity epidemic, fueled by higher and higher average daily calorie counts, continues unabated.

Castelli has good reason to be concerned about cardiovascular disease. He comes from what is now known as a high-risk family, and is the first male member to survive beyond the age of fifty without serious disease. Castelli's father developed atherosclerosis in his legs in his forties, and eventually had to have one amputated. Castelli's older brother had angina. "My mother dropped dead at sixty-four. Most of the members of her family were riddled with heart disease." He was already directing the Framingham Heart Study when he realized his cholesterol level— around 260—was higher than that of most of the heart attack victims in the Study. He lowered his numbers by dramatically changing the way he ate. And he became a runner, until recently, when he has switched to walking. "I did it mostly with diet and exercise, though I have to admit that lately I've had to take medicine. I have terrible genes from my mother's family, but I've outlived my mother and most of my cousins. I ran fifteen to twenty miles a week for twenty years. Then my knee gave out. See, I should have walked so I would still have the knee."

Retired from the Heart Study in 1994, Castelli now teaches and practices what he teaches by running the Framingham Cardiovascular Institute, a clinic where those with risk factors that are difficult to manage can be monitored and taught to change their lifestyle. "I started this clinic after I retired from the Heart Study to try to get at the issue of how do you do this, how do you lower cholesterol? How do you lower blood pressure? How do you get the exercise you need? How do you do diet?" In other words, how could he apply to his own clinical practice what he had been preaching for so long?

Getting the word out for Castelli meant a steep learning curve. He recalls telling doctors about the drugs in development that held the promise of controlling blood cholesterol. "There are twenty-six steps in the manufacturing of cholesterol, and each step has an enzyme with eighteen syllables," he says, mocking his naive attention to medical detail. What he had failed to consider was their end-of-the-day attention span and the aftereffects of the cocktail hour.

"I'm saying all this, working on loading the slides [individually into an old two-inch-by-two-inch projector], my back to these guys, explaining all of the steps. I get close to the end, when I look out at the audience. The entire audience was asleep—including my wife! One guy was snoring. It was a tragic lecture."

But it was important to Castelli to teach others the things that the small cadre of cardiovascular epidemiologists was learning. When he joined the Study in 1965, data were showing that one in eight men between forty and forty-four and one in six men between forty-five and forty-nine would have a heart attack during the early years of follow-up. Two decades after FDR's death, men in their prime were still succumbing. They needed to be informed that they had within them the power to change some of the things that threatened to cut short their lives.

"In Framingham we had church suppers we would be invited to," he says. "We could never say no to anybody. I decided to give a lay lecture. I made up all of these simple little stories about how it all works. I go out there, give this lecture and the people just lit up. The Q and A lasted till midnight." The lesson he learned from lay presentations about conveying simple concepts and speaking in language everyone understands is one he began to apply in his lectures to physicians. It worked then and it still works now.

Today, Castelli knows how to get attention. When he gives a tour of a typical supermarket and packs the ninety minutes with enough nutritional insights to change a diet and a life, I know that I've heard it all before. But I also know that even sophisticated medical students who get the tour can learn simple but important lessons from reading nutrition labels, which Castelli helps them interpret.

Within our supermarkets today are foods that are, in many ways, far better for Americans' health than what was available at the start of the Framingham Heart Study. But the aisles are also filled with temptations, false promises of "lite" and "fat-free" and "enriched." Shelves and shelves of supermarket space are filled with products that threaten health and promote obesity, diabetes, and heart disease. The key to healthy eating is to understand which foods are good choices and which are not.

I have heard Castelli describe his field trips for years, so I had him take me on a private tour of a large Framingham supermarket. Starting at the salad bar, Castelli says that almost nothing there is bad to eat, though he advised skipping the potato salad and walking past the shortcake shells and whipped topping. In addition to the lettuce base, there are about two dozen tubs of mostly healthful additions: tomatoes, cucumbers, green peppers, and onions. "You can even have the eggs, but take the white and leave the yolk," he says. One egg yolk contains 250 milligrams of cholesterol, close to the 300-milligram limit for an entire day. "You could have a lunch with lettuce, tuna chunks, chickpeas, three-bean salad. You could make a delicious lunch." Research has shown that saturated fat in the diet is even worse than dietary cholesterol. So, for salad dressing he advises skipping the bleu cheese and instead drizzling on olive oil or canola oil, rich in monounsaturated fats that can actually lower cholesterol to some degree. Unlike saturated fats and trans fats (hydrogenated to harden at room temperature and prolong shelf life), monounsaturated fat molecules help lower LDL. Monounsaturated fat is also found in nuts, including pecans, almonds, cashews, and walnuts, and in peanuts and avocados. But watch serving sizes, because all fats are high in calories.

At the processed meat case, it's clear that franks have come a long way since German immigrants brought over wieners and wurst. American butchers, looking at what was literally left on the cutting room floor after the steaks, chops, and roasts had been removed from hogs and cows, changed the sausage ingredients. They ground up all the scraps left behind, put them in casings, and called them hot dogs. The American hot dog's history is disputed, as dozens of butchers and immigrants claimed credit. But it's fairly certain that by 1871 Charles Feltman opened the first hot dog stand in Coney Island, serving hot dogs wrapped in bread. Some 130 years later Americans consume 20 billion hot dogs annually.[1] The hot dogs that Framingham participants were eating when the Heart Study began in 1948, along with the rest of American picnickers and baseball fans, typically contained 83.6 percent of their calories in fat—half of it saturated fat. Hebrew National had an advertising campaign claiming that the company had to "answer to a Higher Authority." Castelli scoffs as he remembers it: "You eat those hot dogs, and you're going to *need* a Higher Authority." Today, we can still choose high-fat hot dogs, salami with nearly 90 percent of its calories from fat, and the cold cuts that many Americans call "mystery meat." But Hebrew National, Oscar Mayer, and other meatpackers now offer low-fat hot dogs and cold cuts. Consumers can buy turkey dogs—even tofu dogs—all with lower fat content, and little or no saturated fat. Some hot dogs are fat-free, and we can actually buy fat-free ham.

It's the dairy case that probably shows the greatest change since epidemiology pointed out the value of lowering dietary saturated fat. When Evelyn Langley was encouraging her Framingham neighbors to sign up for the Heart Study, most of those residents were drinking whole milk. Now consumers can wean themselves from whole milk, which is 4 percent fat, to 2 percent milk, on down to 1 percent, and finally to skim—the most heart-healthy choice. Skim milk has all of the calcium and nutrients of whole milk, and by the time most drinkers get accustomed to its taste, a glass of whole milk starts to taste too rich, like drinking straight cream. There is even a half-and-half alternative that is fat-free. Soy drinks are in the dairy case—vanilla or plain for cereal, chocolate for a dessertlike taste—with added calcium and no saturated fat.

Yogurt, something that was scarcely heard of in the United States fifty years ago, now takes up eight, twelve, or more feet in a typical supermarket dairy case. And consumers can choose from dozens of flavors with fat-free options. Again, the labels on dairy products show the amount of saturated fat, the kind to avoid.

Butter is still on the shelves, but the lard that was ubiquitous in the kitchens of our forebears is hard to come by. "Fifty years ago, Crisco was synonymous with hydrogenated fat. Now," Castelli says, holding up a can of the new Crisco and sounding as surprised as anyone, "they use canola oil, soybean oil. It even has a little omega-3 [a protective type of lipid]."

In 1948, butter was the common spread. "Margarine was just coming out," says Castelli. After a period of uncertainty about whether margarine was an improvement over butter, it has become clear that both the saturated fat in butter and the trans fat in margarine are not healthy options. But the shelves now offer better alternatives. The most recent innovation falls into a new category called functional foods, which promise benefits beyond traditional nutrition. Benacol contains a plant derivative which, with one to two tablespoons a day, can actually lower cholesterol levels. Best of all, where once hydrogenated fat dominated, there are now many varieties of olive oil—virgin, extra virgin, light—along with canola oils, rich in healthy monounsaturated fat. Even pan-spritzers like Pam come in olive oil varieties.

But milk and butter aren't the greatest dairy temptations. Ice cream and cheese can quickly pave the road to obesity and heart disease. High-end designer ice cream can have up to twenty grams of fat, most of it saturated, in a single serving. And many people fill their dessert bowls with double the half-cup serving size listed on the label. Today, the frozen food cases give over as much or more room to low-fat ice cream, ice milk, low-fat and fat-free frozen yogurt, frozen soy treats like Tofutti, and a wide variety of fat-free sorbets.

Consumers can buy reduced-fat or fat-free cream cheese; they can get American, cheddar, and provolone cheese with as little as two grams of saturated fat per serving. "Cabot makes a cheddar that's 75 percent fat-free. But you've got to be careful," Castelli says, reading the tiny print on the nutrition label. "An ounce contains 1.5 grams of saturated fat. That's better than the five grams in most cheddars, but it doesn't mean you can eat the entire package." Serving size is everything, and servings should be modest. One can even buy cheese that has no hint of dairy, and no saturated fat, in products like Veggy Slices, often found not with the cheeses but in produce sections, near the tofu. "Sprinkle a little of this on your salad," says Castelli, holding a bottle of Soy Garden parmesan. Or try it on pasta or popcorn—air-popped, or popped in olive or canola oil.

Meat is a problem area, and the biggest problem is quantity. For the past half century, the industry has given Americans tastier and more tender cuts of meat by feeding livestock in ways that add to the fat and marbling. And Americans got used to eating it every day. Still, people can find ground beef ranging from 30 percent fat by weight down to a very lean 5 percent fat. Ground turkey is a low-fat alternative and, in dishes like spaghetti sauce and chili, the taste is nearly indistinguishable. A soy burger, topped with lettuce, tomato, onion, and mustard, can be as tasty as a beef burger with a fraction of the fat. Filet mignon, for those who can afford it, is a reasonable choice for an infrequent meat meal. But with all meats, serving size is key. Americans have gotten used to quarter-pound or larger hamburgers, and steaks and chops that overflow the dinner plate. A healthy serving of meat might be three ounces—about the size of a deck of cards—several times a week. "We eat too much meat in America. In France, they eat these little portions. There's this much room on the side of the plate," Castelli says, indicating about a two-inch border. "They have small portions of delicious food. They'll have a small piece of meat and lots of vegetables." And their heart disease rate is much lower than that of Americans.

Consider the bun you put the meat on, or the bread you eat with it. When it comes to breads, crackers, cereals, and packaged cakes and cookies, Castelli has one rule of thumb: look at the ingredients list, and if the first word isn't *whole,* as in whole wheat or whole grain, don't eat it. Like other cardiovascular disease researchers who have studied nutrition, he is outraged by processed grains, and white flour in particular. "We take perfectly good grain, take out all of the fiber, vitamins, and most of the minerals, and then spray back some of the vitamins," he complains. White flour and white sugar raise triglyceride levels in the blood, and also lower protective HDL levels. A loaf of bread might claim that it is wheat bread, or even boast of being made with seven grains, but if the first ingredient on the label says *enriched flour,* it's not made with much whole wheat flour or any other whole grain. It's made primarily with white flour, and perhaps dusted with a coating of oats or grains. It's difficult to find bakery bread that's made with whole grains, but some packaged breads fill the bill, as long as the first ingredient listed is "whole."

A few years ago, trendy fat-free cookies such as Snackwells caught the nation's attention. Since they were fat-free, the assumption was that they were healthy products. Americans responded by literally waiting for the supermarket delivery trucks, stocking up and often eating entire boxes at a sitting. It didn't take long to find that the cookies, devoid of fat but loaded with enriched flour and sugar, could add to weight problems. They may be fat-free, but they are still high in calories and almost worthless nutritionally.

Most breakfast cereals have the best nutrients removed and large amounts of sugar put in, along with some added vitamins. Whole oat and oat bran cereals such as original Cheerios—not the newer sugar-sweetened varieties—and oat squares are among the healthiest. But hot oatmeal is hard to beat.

A consumer can hardly go wrong with any choice at the fish counter. And like the salad bar, the fish section would not have been anywhere in sight in a grocery store half a century ago. Fish in any form, other than frozen, would have been unheard of in most American inland cities and towns. You'd have had to live near an ocean, or a Great Lake, to get fresh fish. Shrimp has half the saturated fat of the skinless white breast meat of chicken. Just don't cook it with butter and cream. Salmon, tuna, and other fish are fatty, but the omega-3 oils they contain are beneficial.

Today, you can buy hummus, made with chickpeas and tahini, or babaghanoush, made from eggplant and tahini, and use it as a dip for carrots, cocktail tomatoes, or sliced vegetables. You can get strawberries and melons almost year-round. You can become used to the exotic tastes of couscous and tabouli, and learn the pleasures of sushi, available as takeout at many supermarkets. A supermarket in Framingham actually has a sushi chef on the premises. And when you buy packaged and prepared meats, cereals, dairy products, and desserts, they have labels that, by law, tell you what's in them, including the amount of total fat and saturated fat. And soon labels will indicate how much trans fat is contained in foods. The temptations abound, but so do the healthy options.

Castelli's goals are more ambitious than those of the general public. Getting one's total cholesterol, for example, under 200 and LDL cholesterol below 130 is a recommendation of the National Cholesterol Education Program. But 35 percent of heart attacks occur in patients whose total cholesterol is under 200. He urges patients to get their LDL, or bad cholesterol, below 110—much lower if they've already had a heart attack—and to reduce the ratio of total cholesterol to HDL to less than 4.

It's not easy when the things Americans love are full of saturated fats: hamburgers, lamb chops, triple-cream brie, and hot-fudge sundaes. Also ubiquitous are the trans fats, or partially hydrogenated fats. They're abundant in fast foods, margarines, and commercially baked goods. And exercise, which at best burns off about one hundred calories a mile, can't possibly keep up with Big Macs, French fries, and milk shakes. Burning off a fast-food meal could mean about 6.5 hours of exercise.

Despite proselytizing for the heart-healthy gospel, the epidemic of heart disease grows as the American lifestyle spreads. "There are countries where heart attacks, strokes, and atherosclerosis kill hardly anybody. Most of the people who live on this earth never get this disease, but they don't live anywhere near us. They live in Asia and Africa and Latin America. They largely live outside the big cities," says Castelli. But the minute those populations start making money and adopting a Western lifestyle, they begin to get heart disease.

Robert Sullivan remembers hearing Castelli at a dinner meeting of the Middlesex Bar Association. Like all the Framingham participants, Sullivan wasn't medically treated or advised by the Study's doctors. No one ever told him to lose weight, to exercise, to eat vegetables. So that night two decades ago as Sullivan cut his steak, buttered his baked potato before spooning on sour cream, and topped the meal off with pie and ice cream, what Castelli said came as a revelation. "Dr. Castelli spoke about the Study and the findings. He talked about the food we had just eaten. I remember being dumbfounded by his pronounce-ments," says Sullivan.

For Sullivan, the Study was already a family tradition. His father, Walter, and his mother, Katie, had been keeping appointments for thirty years when he heard Castelli's talk. Robert never started smoking, but he liked to eat, and was too busy to exercise. Newspapers had begun publishing results about smoking and about diet and exercise. Sullivan read them, but didn't apply them to his life. Like the old adage in teach-ing that it takes five repetitions to get an idea across, Castelli's speech for Sullivan was the fifth repetition. His words, combined with a belly-ful of fatty foods, made Sullivan think about the consequences of how he was living. Castelli asked if anyone had questions. Sullivan raised his hand and ventured, "What did *you* eat tonight?"

"Not that," said Castelli, pointing to their plates.

"It turned me around. I used to eat like that, meat, potatoes, butter, dessert, all the time. I would eat all of that food and never exercise," Sullivan recalls. "We would get these [Heart Study] questionnaires ask-ing how much, exactly, of each different thing we would eat. *Broccoli?* And I'd think, 'Well . . . never.' The questions started zeroing in on things like carrots, apricots. And I never ate that. Then they'd ask, 'In a given week, how many times do you exercise to the point of working up a sweat?' And it was *never.* I didn't do that."

Now, of course, the news has spread, from the *MetroWest Daily News* to *U.S. News & World Report.* The *New York Times* Science section parses the multiple varieties of fat, discusses hypertension, and illustrates atherosclerotic arteries. The term "heart health" turns up 851 books on Amazon.com. A Google search of the same term reports 2.4 million hits. PBS's *News Hour* partners with the Kaiser Family Foundation to present health news, and often the subject is how to prevent cardiovas-cular disease.

Castelli ends his day by walking down the five flights of stairs he had climbed in the morning. And he continues to teach. "Bend forward. Flex your knees. Walking down is harder on your back and knees than walking up. Walk down with the posture you'd use for walking uphill," he says. It brings to mind the old Marx Brothers "Walk this way" routine—Castelli modeling good downstairs walking posture to guard the back and knees even as the cardiovascular system gets a brief, healthy workout.

SIXTEEN

A Fifty-Year Promise Under Attack

They're not selling *my* data." Evelyn Langley was adamant. In the spring of 2000, she again became a Study activist. She was dead set against the undertaking she read about in a letter to participants and a newspaper report.[1] Boston University had formed a for-profit company called Framingham Genomic Medicine to distribute a half century of data and genetic information from Framingham Heart Study participants.

After attending her twenty-six voluntary appointments over the course of more than half a century, Langley was livid. When she was young, she had walked the streets of Framingham, urging her neighbors to volunteer. And all those years ago, she gave them her word of honor that their privacy would be protected at the same time that the information gleaned from their volunteerism would be available free of charge to scientists. Langley believed that none of the original five thousand participants or their descendants had ever expected their efforts to enrich medical entrepreneurs. Now she was reading and hearing speculation that their personal medical information would be sold, and that a private, for-profit company stood to make money off her and her neighbors' altruism.

More than fifty years after its inception, the Framingham Heart Study ran into a controversy that rivaled the near-shutdown crisis of the late 1960s. It was the worst time of my career, as I struggled to remain publicly neutral and attempted to reassure doubting participants like Langley and scores of others—all the while privately raging that a questionable decision not of my making was threatening the integrity and future of the Study.

I first heard of the proposed venture not in a private conversation or communication, but when I received a copy of the agenda for a September 1999 Boston University faculty retreat. Genetics in Framingham was on the schedule with a presentation by Aram Chobanian, dean of the medical school. He was a mentor during my medical education and early training years, and I continue to respect and admire him, on a professional and personal level, despite our differences on this issue. It was a meeting I nearly missed, and was able to attend only because my plans for an American Heart Association conference in Florida were abruptly canceled by Hurricane Floyd. At the faculty session I learned of the plan to put genetic resources from the Heart Study to an additional use. Boston University was exploring the possibility of establishing a commercial enterprise, later dubbed Framingham Genomic Medicine.

I was shocked. No one had approached me about the idea, and I knew immediately what a powerful impact it would have on the people of Framingham, on the Study's staff, and on our investigators. It was a bold endeavor, poorly conceived and executed. Most worrisome was its insensitivity to the participants' decades of commitment to the Study. My overriding concern was that the citizens of Framingham would perceive that their trust was being violated for profit.

Americans react viscerally when it comes to their DNA. Without enormous amounts of preparatory work, a project aimed at profiting from distributing genetic information was destined to fail, or at the very least provoke a public outcry, because of the sensitivity of the topic.

For years, Fred Ledley, a biotech entrepreneur, had followed the progress of the Framingham Heart Study with a keen interest in the scope and value of its data. When he learned that the university had formed a for-profit company centered on the Study's data, he wanted in. A few months after Framingham Genomic Medicine was incorporated by Boston University officials, Ledley was appointed chief scientific officer. But the advance grassroots education and public relations effort simply wasn't in place when the story was picked up by the media, and a feeding frenzy ensued. Associate Vice President of University Relations David Lampe was called in, belatedly he says, to inform the Study participants. "It was frustrating. What ended up happening was it was misconstrued by the town, the media, and the participants themselves. There was no effort to cash in and make millions. That was not the motivation that I saw from [Boston University] and Ledley. But you're walking on delicate territory when you're talking about people's medical histories and their genome."[2]

Framingham Genomic Medicine was poised to sell data derived from the Study's volunteers. It represented an ethical challenge that never could have been foreseen when the Study began as a publicly funded project of the U.S. Public Health Service and later the National Institutes of Health. By 2000, the public contribution to the Study totaled more than $90 million. Under the new proposal, the data, which had been available to academic researchers at no cost, would be sold to companies willing to pay the price. Later, the university, under pressure from the National Heart, Lung, and Blood Institute, clarified its position: data would still be made available free of charge to academic researchers.

Since the resolution of the 1968 crisis, the Study has been a partnership between the NHLBI and Boston University. Now the university was making arrangements with venture capitalists who raised $21 million for Framingham Genomic Medicine,[3] part of which would be used to create a huge electronic database, one that would be distributed for profit. While not unprecedented—the Human Genome Project combines government-sponsored research with private enterprise—the Study's data are unique and almost entirely paid for by taxpayers' money. Nowhere else is there such a volume of data based on repeated measures from physical examinations and medical histories across three generations. Such public-private partnerships can speed up research and develop treatments and cures that would be slower without the motivation of profit and the participation of big pharmaceutical and biotechnology companies. But profit was not a consideration in the original pact, and privacy was promised forever. We were facing an ethical conundrum.

Framingham Genomic Medicine was banking on a plan for firms to pay annual fees to access data not available elsewhere. The company would have invested millions in packaging data for resale to the pharmaceutical industry. Ledley estimates the cost of the project at $12 million.[4] It would be the first to gain access to such a large, multigenerational archive that includes hundreds of thousands of chest X-rays and electrocardiograms, and hundreds of boxes of paper records documenting half a century of data summarizing dietary and exercise patterns, alcohol and cigarette use, even information not previously mined. The archive also contains thousands of frozen DNA samples from Framingham volunteers, from which genetic data could be derived. Far better than most epidemiologic data, which depend on recall and old records, Framingham has detailed participant information, including sensitive clinical and behavioral data. Some of it is still not part of a computer database, its secrets locked within boxes of stored paper records.

Sorting it all out would fill a void for drug companies, especially if they could have access to DNA samples or derived genetic data, allowing them to make a connection between specific genes and health outcomes. Connecting clinical data with genetic test results could lead to important discoveries and possibly new treatments. Ledley argued in a *Boston Globe* opinion article: "Scientists alone cannot put a human face on genomics. This will require the involvement of people who are willing to participate in clinical research that compares each individual's personal experience of health and disease with the chemical structure of their genes."[5]

But the business plan had less to do with creating a database for the benefit of the Framingham Heart Study and more to do with establishing a marketable product. And as the situation unfolded, it became clear that the half century of trust built between the Study and the community was in jeopardy.

Arthur Caplan is a son of Framingham. He grew up there, went to Framingham North High School, and worked at Pinefield Pharmacy, his father's store. As a professor of medical ethics at the University of Pennsylvania, he lectures students about aspects of the Heart Study and knows as only a local can the respect it garners. His parents arrived in town in 1957, so he and his family were never participants, but the town and the Study will always be part of him. He heard the urban legends surrounding the Study. "They take your body when you die," kids would whisper (not true).[6] As an adolescent cashier in his father's drugstore, he figures he sold a lot of life-shortening cigarettes to Framingham residents before they helped prove that smoking led to heart disease.

Residents found out about Framingham Genomic Medicine from a brief letter they received from Boston University announcing the establishment of a company. While BU had partnered with the NHLBI in overseeing many aspects of the Heart Study for thirty years at that point, the participants were accustomed to receiving news about the Study on NHLBI's Framingham Heart Study letterhead.

The BU letter provided little detail of the nature of the venture, who was promoting the plan, and how it might affect them. It didn't even mention the name of the company or that participants' data would be used as a commodity:

April 27, 2000

Dear Friend of Framingham or Study Participant:

The Framingham Study is entering an important new era of medical research which involves linking the discoveries in genetics to advances in medicine. Boston University has recently formed a company to expand research and pursue genetic discoveries. The Company will work with academic investigators and pharmaceutical companies to apply the discoveries from the Framingham Study to advance the development of new drugs and diagnostic products and also to develop new approaches to medical care.

There will be no changes in the conduct of the study or in your visits to the clinic. Everyone associated with Boston University and the Company is mindful of the extraordinary contributions you have made to advance medical science. The Company understands its obligation to maintain exemplary ethical standards and will be establishing an independent ethical review group to monitor the Company's activities. The Company will be located in Framingham and will work with the Friends of the Framingham Heart Study to direct a portion of its resources back to the community and the Study. We will keep you informed about this activity as it progresses over the next several months. Should you have any questions, please don't hesitate to call [800 number]. As always, we appreciate your continued dedication and support to the Study.

Yours sincerely,

William Castelli, M.D.

Philip A. Wolf, M.D.
Principal Investigator[7]

Caplan was called in to look into the complex moral issues involved, but only after the town's residents were already roiled. "Of all the things that ever happened to the Study, the near closing was the biggest threat. The second biggest was Framingham Genomics. By the time I got there, the house was on fire," he says. Among his strongest pieces of advice was that volunteers be thoroughly informed about exactly how the profits from the new company would benefit them and the Study. That question was never answered to their satisfaction, or to mine.

Castelli was well known to Study participants. It was for this very reason that Ledley and officials at Boston University sought his endorsement of the plan and his signature on the letter. He believed the venture offered an opportunity to use Study data to promote gene discovery and develop new treatments. But he didn't anticipate the backlash that would result from lending his name to the cause. "I think the press riled people up," says Castelli, adding that he still felt that such a venture would have been beneficial to science.[8]

The news came as a fait accompli on Boston University's letterhead. "That was a big mistake. No one recognizes BU letterhead. They [Study participants] all recognize NIH letterhead," says Ledley. But using the federal government's letterhead was out of the question. Claude Lenfant, director of the NHLBI, didn't want any reference to the Heart Institute mentioned in the letter because he did not want to give the appearance that NHLBI endorsed the project.

The media were having a field day with the story. The community learned more about Framingham Genomic Medicine through reports in the *Boston Business Journal*, the *Boston Globe*, and *MetroWest Daily*, which ran, among other items, an editorial cartoon. It showed a Framingham Heart Study volunteer next to his mailbox, prone and clutching his heart. In his hand was a letter saying "Data for profit scheme."[9]

What the participants read was that Framingham Genomic Medicine and Boston University would profit from their decades of effort. "The most lethal agent you can apply to altruism is money," says Caplan. "They were immediately up in arms. They'd say, 'Why are they taking our gift and trying to profit?'"

Participants began calling the Study, asking for me, coordinators, anyone they could get on the phone. For the most part, they were irate. Again and again I read the word "betrayal" in their letters, heard the hurt and mistrust in their voices. Confusion gave rise to panic. A few people began to talk about boycotting the Study. Even loyal participants like Evelyn Langley threatened to join a boycott and drop out.

Letters to the Editor columns became a forum. Study participant Sandra Fitts wrote a supportive letter to the local newspaper to say that the new plan could lead to more rapid discovery of cures for heart disease.[10]

A letter to *MetroWest Daily* from participant Daniel J. Liberatore closed with the sentiment that I saw as a looming disaster: "The financial gains from this enterprise undoubtedly will be enormous. However, the precedent set may hamper medical research for decades to come. The betrayal of trust is not a trivial matter. This will become evident as others come to realize just what has happened to their sensitive medical data. Personally, I will no longer participate in the Framingham Heart Study or any other. Had I the slightest inkling of what was to be, I would never have enrolled. I now bitterly regret having done so."[11] After the failure of the venture, Liberatore was satisfied that the right thing was done and his privacy would continue to be protected. He remains an active participant in the Study.

Although I wasn't involved in Framingham Genomic Medicine, I recognized the effect it was having on participants and struggled to reduce the damage. In a letter to 5,500 participants, I let them know that whatever the fate of Framingham Genomic Medicine, we would safeguard their data, their privacy, and their interests. I invited them to openly express their views on the venture, which I would share with the university. The responses were overwhelmingly negative.

"We would rather see Framingham Heart Study information and data destroyed than have it turned over to a 'for-profit' entity."

"If this change takes place, it will be the demise of the program as we know it. So many people have put time and effort into this project. It seems a shame it all comes down to MONEY."

"My family was in the original Study in the 1940s. . . . To think that this commitment, free of money-profit, could be used for others to directly profit financially from, in any manner, is truly an exploitation."

"Money and profit are necessary ingredients for any company; however, we do not think that the Framingham Heart Study was started with any of these thoughts in mind nor were they in the hearts of the many people who have so generously given their time and their physical capacities to research. . . . Please don't let years of dedication and loyalty end."

"This is a brave new world territory, and despite oversight committees, policies, and regulations, these safeguards are only as good as the individuals who actually handle the data. . . . I would like to keep an open mind on what is about to unfold, although at first glance, it's an effort."

"The Study was initiated to help all mankind, *not* for financial profit to any company or individual. We stand by this principle and will not sign any consent form for use of this data for profit."

Through the first half of 2000, we held staff meetings aimed at educating everyone about how to answer the endless questions we were getting from participants during routine clinic exams. We knew the Study participants would need a crash course in genetics and in how their DNA and data would be used in research. Some thought the very name of the company, Framingham Genomic Medicine, was troublesome. It seemed to link the venture with the venerated Study in a way that was simply not true. And the word "genomics" itself triggers confusion and even alarm over the kinds of information genes can reveal.

A few volunteers refused to sign the usual consent forms during their clinic exams or called to withdraw their prior consent to distribute data or DNA. Their main questions concerned privacy and confidentiality, but they also wanted to know details of the financial arrangements of the company: who would make money, how profits would be used, and whether there would be any conflicts of interest. Caplan points out that if this venture was designed to benefit the Study and the participants, as the university and Ledley claimed in newspaper articles, there would have been a set of bylaws establishing how corporate profits would be used for those purposes. Nothing of the sort was ever put in writing.

The business questions were ones I could not answer then, nor can I now. I know that Ledley asked me in the fall of 1999 if I would be interested in serving as a consultant for the company. I said no. Concerned about conflicts of interest, I asked if other investigators were being offered consultancies or royalties from the venture. He explained that incentives to investigators would be consistent with existing regulations and university policies.

The company, perhaps surprised by the outrage of the participants, responded with a plan. It would pay for a science education program in local schools and set up a fund to benefit the town. It would donate computers. It would give the Friends of the Framingham Heart Study founders' shares in the company.[12] But these gestures were perceived as an attempt to buy goodwill. To many in the Study, their voluntary contributions were being turned into a commodity for sale.

Confronted with this rift in trust, we had to protect the integrity of the Study. That's when Caplan came in, with the clear mission of looking into the ethics of the deal and representing the interests of the participants. He talked with me, with participants, with representatives of the university and of the NHLBI. He joined us as we convened a town forum in August 2000 attended by about two hundred highly vocal volunteers.

Jay Lander, vice president of the participant group, Friends of the Framingham Heart Study, read from a letter he had sent to key decision makers.

On behalf of the participants of the Framingham Heart Study, we, the Board of the Friends of the Framingham Heart Study, wish to advise you of the many concerns that have been expressed to us by the Study participants regarding the proposed agreement between Boston University and Framingham Genomic Medicine to utilize Heart Study data for genomic research for profit. It is our understanding that the parties that are negotiating this proposed venture recognize the importance of preserving participant confidentiality, ensuring equal access to all of the Heart Study data that would be shared with the new company, and avoiding actual or perceived conflicts of interest. We anticipate that the leadership of the National Heart, Lung, and Blood Institute will insist that the above principles are adhered to, and we are in agreement with these goals. Nevertheless, the initial overwhelming reaction of a significant number of participants, as expressed to members of our board as well as to staff of the Study, has been negative. A common theme is one of suspicion and a feeling of betrayal of the original altruistic motives of the Study and its participants by the proposed commercialization of the Study data.

It is our belief that there must be a close ethical oversight of any new venture, which holds great promise for new drug development, but also raises important concerns. Our over-riding goal is to protect and maintain the integrity of the Framingham Heart Study, and to promote its scientific mission. It is absolutely essential that the work of the Framingham Heart Study be allowed to continue. This can only be accomplished with the continued support of the participants. We feel that the data collected by the Study over the past fifty years, if handled properly, has the potential to save the lives of our grandchildren and their grandchildren. To this end, we ask you to take the community's concerns to heart as you negotiate the details of this venture, and to reassure the participants by community education and dialogue as soon as possible in order to restore their confidence in your commitment to preserve the integrity of this Study.[13]

The uproar was about ethics and fifty years of mutual trust. "Some of the concessions—building clinics, creating a fund the people of Framingham could control, agreeing that any discoveries could benefit poor countries at no cost—came too late," says Caplan. "The long and the short of it is that because the company didn't pay attention to the ethics of what this Study was about, they couldn't do it. The people were mad and the investors got nervous."

By the autumn of 2000, Boston University, wisely but reluctantly, pulled the plug on the project. To those of us who had spent our careers working with Study participants the outcome of this venture was predictable.

The story of Framingham Genomic Medicine offers a lesson Caplan is teaching his students. In fact, he and I are collaborating on an article that may help other research groups avoid similar mishaps. Perhaps the for-profit plan met the letter of the law. But bypassing the consideration of participants spelled disaster. In the end, the perceptions of the community that their altruism was for sale could not be overcome.

During the three years that followed Framingham Genomic Medicine, we worked to solidify our bond with the participants and reestablish a climate of mutual trust. We also had fences to mend with the scientific community, which was concerned that Heart Study data were not freely available to outside researchers. To address those concerns, the Study expanded and publicized a system to share our data and DNA samples, free of charge, with scientists. The Framingham Genomic Medicine idea had apparently been formulated within the university's technology transfer office by people who were familiar with how other institutions had made money by patenting discoveries. This was an attempt to do the same at Boston University—a business proposal that was engineered by individuals unfamiliar with the decades of Framingham altruism.

The value of the Study is so much greater than any amount of income that would have flowed into the university from this venture. Framingham Genomic Medicine came close to disturbing the delicate balance of trust that has been the secret of our success for over fifty years. Millions of dollars might have been made, but if the second generation of volunteers had stopped coming in, or a third generation had refused to sign up, the Study and all it had yet to uncover would have been killed. The flow of data would have dried up and the Study would have ended.

The predicament raised important issues of how private industry can work with information generated from tax dollars. Those issues remain unresolved.

Dawber and Kannel lived through their low point with the Study when it almost closed. This was mine. And just as the Study rebounded from its near shutdown in the late 1960s with the recruitment of the second generation of participants, so too have we recovered from another crisis, only to emerge with a renewed mission.

SEVENTEEN

The Genetic Revolution

The launch of the Third Generation Study in April of 2002 was our most ambitious undertaking since we enrolled a second generation of participants in 1971. Much of our research focus today is directed at identifying genetic causes of cardiovascular disease and the risk factors that we have been examining since the Study's inception: high blood pressure, high cholesterol, low HDL, diabetes, and obesity, as well as many other common traits and disorders. This new recruitment phase will give us an additional generation of participants with DNA that will be paired with three full generations of clinical and laboratory data. That combination of information will enhance what we, and the rest of the scientific community, discover in the future. Anything that we measure about participants can also be studied genetically.

The Heart Study has health measures from grandparents when they were in their thirties and beyond, and identical data from their children at the same age, and now we are collecting the same information from their grandchildren at the same age. There is no missing link. It is a perfect and unparalleled design for genetic research.

Genetic research at Framingham may, at first blush, appear to be a contradiction. We have spent half a century pursuing risk factors: high blood pressure, high levels of cholesterol, diet, obesity, smoking, and a sedentary lifestyle. And we have insisted on the importance of modifying those risk factors as a means to reduce the threat of heart disease and stroke. So how do genes enter the picture? If heart disease risk is in our genes, why worry about lifestyle?

All the traditional risk factors, those that are modifiable and even those no one can control, such as age and sex, don't account for all the occurrences of heart disease. While the percentage of disease explained by risk factors is somewhat controversial, it is probably about half. It's the other half we're beginning to go after with genetic research. There are dozens—even hundreds—of genes involved in heart disease, and it is very likely that some of those genes act to accelerate cardiovascular disease susceptibility in people who smoke, eat a high-saturated-fat diet, fail to exercise, or otherwise ignore lifestyle risks. Conversely, a genetic predisposition to heart disease may not become manifest in someone with a low burden of risk factors—a nonsmoker with low blood pressure, a low LDL cholesterol level, and a high level of HDL. There is also a complex interaction among traditional risk factors, lifestyle choices, and disease-susceptibility genes. Some genes contribute to patterns of established risk factors such as high blood pressure, or even to vulnerability to complications of cigarette smoking. And genes for heart disease are not exclusive; they act in concert and their effects can be synergistic or additive. In searching for heart disease genes we may discover novel risk factors in critical cardiovascular disease pathways related to atherosclerosis, plaque vulnerability, blood clotting, and inflammation. These discoveries could lead to new treatments for people with heart disease and approaches to its prevention.

To pave the way for future genetic research we began collecting blood samples for DNA extraction in the early 1990s. The first generation of participants by then was a skewed sample. Those most vulnerable to heart disease were already dead. Those still alive were the healthy survivors. But the DNA gathered from their children in the Framingham Offspring Study is more representative of the general population. And with the enrollment of 3,500 of the grandchildren of the original volunteers for the Third Generation Study, the DNA resources available for research will grow vastly.

In addition to collecting more than eight thousand blood samples for DNA, we have also been accumulating an unprecedented store of immortalized cell lines for genetic research and now have more than five thousand such samples. These are frozen and can later be grown in cell culture dishes to create an unlimited amount of DNA that will survive for research use after the participants are no longer alive.

Framingham DNA samples and cell lines are essential for genetic research. But major discoveries will not happen at a sufficiently fast pace unless we actively seek out collaborations with experts and give them full access to Framingham's data and DNA resources. In keeping with the original promise of the Study for worldwide benefit, any nonprofit academic institution can send us an application and obtain data and DNA samples. An independent committee of outside researchers reviews the application and oversees approval. Applicants agree to abide by the informed consent provided by the participants and pledge to maintain the confidentiality of those who contributed the biological specimens. This enables scientists to perform myriad tests in an effort to discover genes contributing to the common conditions found in the Study's participants—hypertension, high levels of cholesterol, obesity, diabetes, asthma, and even noncardiac conditions like arthritis, osteoporosis, and Alzheimer's disease.

Obesity is a good example of a common and risky condition for which genetic research is likely to yield new insights and treatments. The epidemic of overweight and obesity threatens to reverse many of the advances made thus far in heart disease prevention. At nearly every Framingham Heart Study clinic examination, we have taken height and weight and calculated body mass index, the measurement that defines overweight and obesity. With DNA samples on multiple generations of family members, we can search for genes that may be responsible for the condition. One approach is a "genome scan" in which 400 genetic markers, representing chromosomal regions across the entire human genome, are genotyped—characterized for subtle variation—in each of 1,800 participants from the largest families in the study. So far, we've completed 750,000 genotyping tests in the first two Study generations in search of genetic causes of various diseases and traits; in the very near future the analysis of 1.5 million additional genetic markers in the third generation will refine our quest for disease-promoting genes.

Within families, if there is a gene in a particular region of a chromosome that causes a disease such as obesity, the DNA of relatives who are obese will reveal the same variant at markers adjacent to the causal gene, while their thin relatives will have different variants at those neighboring locations. Analyzing these patterns across our families will tell us if a chromosomal region is "linked" to obesity. These methods allow us to narrow the search for disease-promoting genes. Nevertheless, the regions spanning linkage peaks typically are vast: 20 million to 40 million DNA bases out of 3 billion in the entire genome. That still leaves us looking for a needle in a haystack—but we're no longer looking for a needle on a continent.

And we're making progress. For obesity, we have found a region with strong evidence of linkage, something we can pursue further by studying individual genes within that region.[1] The effort of identifying individual genes causing obesity, however, will be huge. If our scientists are lucky, they may succeed in taking the linkage results to the next step. More probably, on the basis of published Framingham results, researchers at other institutions will once again receive the baton, follow up on our clues, and advance the cause of gene discovery.

Finding genetic causes of obesity will not be likely to lead to gene replacement therapy, where altered genes are introduced into people with the defects. Rather, understanding genetic contributions to overweight and obesity will identify new therapeutic targets for designer drugs to block hunger, regulate appetite, or increase the feeling of satiety.

Obesity is but one example of genetic research the study can do remarkably well because the condition is so common. The same is true of high blood pressure, which fully a quarter of our participants have. Our genome scan results located a region on chromosome 17 linked to blood pressure patterns within families.[2] We don't possess the expertise to pursue this finding on our own, so we are collaborating with Richard Lifton at Yale University to search for genes that affect blood pressure. Just as Michael Brown and Joseph Goldstein looked at families that had the rare LDL cholesterol receptor mutations causing extraordinarily high cholesterol levels, Lifton is analyzing families with unusual genetic defects causing high or low blood pressure. Unfortunately, this is a difficult task; the search for hypertension genes has frustrated researchers for many years. There could be well over a hundred genes responsible for hypertension, each contributing in different families. But perhaps some day medications will be designed to target these genetic defects and the pathways they represent, reducing the hit-and-miss prescribing that physicians do now as they try to find the right medication for lowering blood pressure. Research will also sort out the genetic variations responsible for how human beings handle salt. In some, it raises blood pressure, while others have less to fear.

Part of the excitement in studying genetic factors is the opportunity to identify novel mechanisms of risk. In a recent project, we examined variants in an estrogen receptor gene.[3] We looked at building-block alterations in the structure of the DNA for the estrogen receptor gene. We found a particular variant in that receptor that was present in two doses (two defective copies) in 20 percent of our Study population. And that variation carried with it a threefold increase in risk for heart attack. What was even more surprising than the fact that it was relatively common was that the risk was increased in men. We don't often think about men and estrogen in the same sentence, but men, too, have circulating estrogen, though at far lower levels than women. This finding suggested that estrogen receptor variation may be an important contributor to cardiovascular disease risk. The result could advance the understanding of why some women given estrogen replacement are actually at increased risk for heart disease. Further research might help scientists determine which women can safely receive estrogen replacement therapy to help protect them from osteoporosis without putting them at increased risk for heart disease.

Because we freely share our data and DNA with the scientific community, Framingham is poised at the cutting edge of the genetic revolution. We hope to continue making discoveries that will change the practice of medicine.

Epilogue

Franklin Delano Roosevelt, a victim of the medical ignorance of his time, died in 1945 from complications of uncontrolled hypertension.

Fifty-five years later, Dick Cheney was elected vice president of the United States after suffering three heart attacks. His first was in 1978, his second in 1984. After a third in 1988, he had quadruple bypass surgery.

In 2001, with the voters fully aware of his medical history, he took office, a figurative heartbeat away from the presidency. His cardiologists assured the public that Cheney's heart disease was controllable. That year, after he had what his doctors characterized as a "very slight" heart attack—his fourth—he underwent angioplasty and received an implanted stent to help keep a blocked artery open. That's when Cheney said he quit smoking, modified his diet, and started thirty-minute daily workouts on a treadmill.[1]

In June 2001, Cheney wore a monitoring device for a weekend, one so small and inconspicuous that he could go about his usual business. The miniature electrocardiographic device showed evidence of arrhythmia; his heart sped up abnormally four times, to 135 beats per minute. Doctors suggested he should join 52,000 other Americans who have cardioverter-defibrillators implanted in their chests to protect their hearts from irregular rhythms.[2] The device is ever on the alert, like an on-call physician who needs no sleep, to deliver an electrical burst to shock the heart back into a normal rhythm.

Cheney's ability to hold office can be attributed to the miraculous advances in treating heart disease after it takes hold. His case exemplifies the success of the revolution in high-technology heart disease treatment, but it also tells of a failure to reap the benefits of the prevention revolution. Angioplasty can alter the internal geometry of the arteries, but it does nothing to change the biology of the disease. Following life-threatening events, Cheney's lifestyle adjustments can be attributed to an understanding of risk factors and the tools to control them. Without such adjustments and medication, recurrence is almost inevitable.

But heart disease continues into the twenty-first century as the leading cause of death in the United States. It is this country's most costly medical condition. The price tag for treating heart disease at the turn of the century is nearly $330 billion per year, and 62 million Americans suffer from it. In the United States, someone dies of cardiovascular disease every thirty-three seconds.[3]

The skill and technology now available to save patients with heart disease are miraculous—and costly. Preventing diseases in the first place is a far better option. Yet many Americans ignore the scientifically proven advice to eat a balanced, low-cholesterol, low-saturated-fat diet, exercise, avoid cigarettes, and control their blood pressure and cholesterol numbers, taking medication if necessary to reduce levels. Too many are passively waiting for a magic pill.

Could there be a magic pill? A genetic understanding of HDL cholesterol, the protective cholesterol, for example, could enable scientists to develop treatment to raise HDL and provide protection from heart disease. At this point, it may seem like science fiction, but such drugs are in development.

We already know that lifestyle measures and attention to medical advice on controlling blood pressure and lowering cholesterol can prevent heart disease and stroke and add years of quality life. That is a lot easier than having to enter the arena of the second heart revolution— angioplasty and coronary bypass surgery. In the future, the advice medicine has to give will be refined. Research will reveal additional risk factors and new treatments for those we already understand. It recently was revealed that cholesterol levels in patients with heart disease can and should be reduced to levels unheard of outside of rural China. The science of genetics will uncover susceptibility genes, alerting some individuals to their increased risk in time to take lifesaving steps. But even now, the simple advice works. Medical science, with considerable help from the Framingham Heart Study, has brought us to a point where we have enormous control over our risk factors. We can change the natural history of our number one killer—even those of us who chose our parents unwisely.

NOTES

INTRODUCTION

1. U.S. Census, *American Community Survey Profile*, Table 2, 2002.

2. Statement of Claude Lenfant, Director, National Heart, Lung, and Blood Institute, February 17, 2000.

3. Laurie G. Futterman and Louis Lemberg, "The Framingham Heart Study: A Pivotal Legacy of the Last Millennium," *American Journal of Critical Care*, March 2000.

4. National Institutes of Health, National Heart, Lung, and Blood Institute, Framingham Heart Study Archives.

ONE: A KILLER OF PAUPERS AND PRESIDENTS

1. Hugh Gregory Gallagher, *FDR's Splendid Deception* (New York: Dodd, Mead, 1985), p. 163.

2. Doris Kearns Goodwin, *No Ordinary Time* (New York: Simon and Schuster, 1994), pp. 586–87.

3. John Bumgarner, *The Health of Presidents* (Jefferson, N.C.: McFarland, 1994), p. 211.

4. *Washington Post*, April 13, 1945.

5. Bumgarner, *The Health of Presidents*, p. 211.

6. Franklin Delano Roosevelt, Second Inaugural Address, January 20, 1937.

7. Franklin Delano Roosevelt, First Inaugural Address, March 4, 1933.

8. H. G. Bruenn, "Clinical Notes on the Illness and Death of President Franklin D. Roosevelt," *Annals of Internal Medicine*, March 1970.

9. Bumgarner, *The Health of Presidents*, p. 216.

10. Ibid., p. 215.

11. Ibid.

12. Goodwin, *No Ordinary Time*, p. 491.

13. Franz H. Messerli, "This Day 50 Years Ago," *New England Journal of Medicine,* April 13, 1995.

14. Goodwin, *No Ordinary Time,* p. 499.

15. Bumgarner, *The Health of Presidents,* p. 212.

16. Goodwin, *No Ordinary Time,* p. 493.

17. Ray W. Gifford Jr., "FDR and Hypertension: If We'd Only Known Then What We Know Now," *Geriatrics,* January 1996.

18. Bumgarner, *The Health of Presidents,* p. 214.

19. Goodwin, *No Ordinary Time,* p. 497.

20. Bumgarner, *The Health of Presidents,* p. 214.

21. Bruenn, "Clinical Notes."

22. Gifford, "FDR and Hypertension."

23. Bruenn, "Clinical Notes."

24. Bumgarner, *The Health of Presidents,* p. 217.

25. Gallagher, *FDR's Splendid Deception,* p. 197.

26. Franklin Delano Roosevelt, Fourth Inaugural Address, January 20, 1945.

27. Frances Perkins, *The Roosevelt I Knew* (New York: Viking Press, 1946), p. 393.

28. Goodwin, *No Ordinary Time,* p. 585.

29. Bruenn, "Clinical Notes."

30. Bernard Asbell, *When F.D.R. Died* (New York: Holt, Rinehart and Winston, 1961), p. 20.

31. Ibid., pp. 22–25.

32. Ibid., p. 15.

33. Ibid., p. 22.

34. Goodwin, *No Ordinary Time,* p. 602.

35. Asbell, *When F.D.R. Died,* p. 38.

36. Ibid., p. 128.

37. Press reports, April 13, 1945.

38. Ibid. Whether Eleanor Roosevelt said "slipped away" or "slept away" remains uncertain, owing to the vagaries of the dictation and teletype processes of the time.

39. Ibid.

40. Interview with William Walenski, spring 2003.

41. O. F. Hedley, "Studies of Heart Disease Mortality," *Public Health Bulletin,* No. 231, October 1936.

TWO: THE DAWN OF PEACE AND PROSPERITY— AND A DEADLY LIFESTYLE

1. *This Fabulous Century: 1940–1950,* series editor, Ezra Brown, (New York: Time-Life Books, 1985), p. 210.

2. Ibid., p. 330.

3. Ibid., p. 210.

4. Nirav J. Mehta and Ijaz A. Khan, "Cardiology's 10 Greatest Discoveries of the 20th Century," *Texas Heart Institute Journal,* November 3, 2002.

5. Interview for *U.S. News & World Report,* September 7, 1998.

6. U.S. Department of Health and Human Services publication, "Percentage of Current Cigarette Smokers Among Adults, by Year, United States, 1944–1986."

7. *This Fabulous Century,* p. 215.

8. Research documents, Framingham Historical Society.

9. *This Fabulous Century,* p. 127.

10. Louis I. Dublin and Mortimer Speigelman, "Factors in the Higher Mortality of Our Older Age Group," *American Journal of Public Health,* April 1952.

11. www.ricedietprogram.com.

12. Franz J. Ingelfinger and Arnold S. Relman, eds., *Controversy in Internal Medicine* (Philadelphia: Saunders, 1966).

13. Interviews, January and February 2001.

14. Interview, December 2000.

15. Interviews, 1998–2003.

16. Victoria A. Harden, NIH Historian, "A Short History of the National Institutes of Health," http://history.nih.gov/exhibits/history/full-text.html.

17. Vannevar Bush, "Science: The Endless Frontier," Report to the President from the Director of the Office of Scientific Research and Development, July 1945.

18. National Heart, Lung, and Blood Institute, Framingham Heart Study Archives.

19. Correspondence, American College of Cardiology, 2002.

20. Howard A. Rusk, "U.S. Fight on Heart Disease Backed Widely by Citizens." *New York Times,* July 11, 1948.

THREE: GATHERING EVIDENCE, BUILDING ON CLUES

1. Interviews, 1998–2003.

2. William Heberden, *Commentary on the History and Cure of Diseases* (London: T. Payne, 1802).

3. National Office of Vital Statistics, December 1947.

4. National Center for Health Statistics, 2003 telephone call.

5. Paul D. White, "Heart Disease Forty Years Ago and Now," paper read at the fifth clinical session of the American Medical Association, December 1951.

6. Ibid.

7. James B. Herrick, "Clinical Features of the Coronary Arteries," *Journal of the American Medical Association,* December 7, 1942.

8. James B. Herrick, "An Intimate Account of My Early Experience with Coronary Thrombosis," *American Heart Journal,* January 1944.

9. Ibid.

10. James B. Herrick, "Thrombosis of the Coronary Arteries," *Journal of the American Medical Association,* 59:2015–20, 1919.

11. Herrick, "An Intimate Account."

12. Julius H. Comroe, Jr., *Exploring the Heart* (Toronto: George J. McLeod, 1983), p. 118.

13. Rene Favaloro, "A Revival of Paul Dudley White: An Overview of Present Medical Practice and of Our Society," *Circulation,* March 30, 1999.

14. Personal interview with William Zukel, March 21 and 25, 2003.

15. Donald S. Fredrickson, "Phenotyping. On Reaching Base Camp," *Circulation,* April 1993.

16. Ancel Keys, *Coronary Heart Disease in Seven Countries,* American Heart Association Monograph No. 29, 1970.

17. Paul Dudley White, *Heart Disease* (New York: Macmillan, 1931).

18. Comroe, *Exploring the Heart,* pp. 152–53.

19. Ibid., p. 234.

20. Ibid., pp. 233–35.

21. Menard Gertler and Paul D. White, *Coronary Heart Disease in Young Adults* (Cambridge: Harvard University Press, 1954), p. 8.

22. Donald Gasner, Elliott H. McCleary, The American Medical Association Straight-Talk, No-Nonsense Guide to Heart Care, p. 3 (New York: Random House, 1984).

23. Interviews, 1998–2003.

FOUR: A STRUGGLE FOR IDENTITY

1. National Heart, Lung, and Blood Institute, Framingham Heart Study Archives.

2. Interviews, 1998–2003.

3. NHLBI, Framingham Archives.

4. Research documents, Framingham Historical Society.

5. Interviews, March 21 and 25, 2003.

6. Interviews, 1998–2003.

7. Research documents, Framingham Historical Society.

8. Interview for *U.S. News & World Report,* September 7, 1998.

9. Framingham Historical Society, Exhibition: "Zeal for Healing: A Framingham Trait," April 21–August 18, 2001.

10. Framingham Historical Society.

11. NHLBI, Framingham Archives.

12. Gerald M. Oppenheimer, "Becoming the Framingham Study: 1947–1950," unpublished, March 2004.

13. Ibid.

14. Ibid.

15. Interview with William Zukel, March 2003.

16. NHLBI, Framingham Archives.

FIVE: THE PEOPLE WHO CHANGED AMERICA'S HEART: VOICES FROM FRAMINGHAM

1. Statement of plan, November 1, 1949, NHLBI, Framingham Archives.
2. Gloria Negri, "Anna Skinner, 105, Heart Study Volunteer," *Boston Globe,* March 29, 2003.

SIX: THE LAUNCH OF A GOLD STANDARD

1. Interviews with Thomas Roy Dawber between April 2000 and August 2003.
2. Notation, NHLBI Framingham Archives.
3. Gerald M. Oppenheimer, "Becoming the Framingham Study: 1947–1950," unpublished, March 2004.
4. Ibid.
5. Ibid.
6. Ibid.
7. Form letters, NHLBI, Framingham Archives.
8. Examination questionnaire, NHLBI, Framingham Archives.
9. Interview with Patricia McNamara, April 2001.
10. 1948–49 documents, NHLBI, Framingham Archives.
11. Paul D. White, "Heart Disease Forty Years Ago and Now," read at the Fifth Clinical Session of the American Medical Association, December 8, 1951.
12. Paul Oglesby, *Take Heart: The Life and Prescription for Living of Paul Dudley White, the World's Premier Cardiologist* (Cambridge: Harvard University Press, 1986).
13. Rene Favaloro, "A Revival of Paul Dudley White: An Overview of Present Medical Practice and of Our Society," *Circulation,* March 30, 1999.

SEVEN: WRESTING CONTROL FROM FATE: RESULTS THAT CHANGED A CULTURE

1. Interview with Brian King, February 2002.
2. Interview with Kitty Walsh, April 2002.
3. Thomas R. Dawber, Felix E. Moore, and George V. Mann, "Coronary Heart Disease in the Framingham Study," *American Journal of Public Health,* April 1957.
4. W. B. Kannel et al., "Epidemiologic Assessment of the Role of Blood Pressure in Stroke: The Framingham Heart Study," *Journal of the American Medical Association,* October 12, 1970.
5. Interview with Henry Blackburn, March 30, 2003.
6. 1956 notation, NHLBI, Framingham Archives.
7. Thomas D. Dublin, NHI, report on visit to Framingham, March 1956, NHLBI, Framingham Archives.

8. Ibid.

9. Interview with Jeremiah Stamler, March 25, 2002.

10. W. B. Kannel et al., "Factors of Risk in the Development of Coronary Heart Disease—Six-Year Follow-Up Experience," *Annals of Internal Medicine*, July 1961.

11. Ibid.

12. Ibid.

13. Thomas R. Dawber, memo to staff, December 1954, NHLBI, Framingham Archives.

EIGHT: A NEAR-DEATH EXPERIENCE

1. Nirav J. Mehta and Ijaz Khan, "Cardiology's 10 Greatest Discoveries of the 20th Century," *Texas Heart Institute Journal*, November 3, 2002.

2. Thomas Royle Dawber, *The Framingham Study* (Cambridge, Mass.: Commonwealth Fund, 1980) p. 27.

3. Interview with William Zukel, March 2003.

4. Interview with Manning Feinleib, March 2003.

5. Site visit report, 1965, NHLBI, Framingham Archives.

6. Abraham Lilienfeld, memo of recommendations, November 1966, NHLBI, Framingham Archives.

7. Ibid.

8. Thomas Dawber, transcript of speech, Boston University, September 1988.

9. William Zukel, memo to Theodore Cooper, May 28, 1969, NHLBI, Framingham Archives.

10. Framingham Heart Study Advisory Committee Report, June 16, 1969, NHLBI, Framingham Archives.

11. "Framingham Heart Study Cut," *Drug Research Reports*, October 15, 1969, NHLBI, Framingham Archives.

12. "Operational Steps for Close-Out of Framingham," August 1969, NHLBI, Framingham Archives.

13. Senator William Proxmire, "Another Example of Misplaced Priorities," *Congressional Record*, October 3, 1969.

14. The letters from Dr. White and President Nixon are reprinted courtesy of Jack Eckert, reference librarian, Francis A. Countway Library of Medicine, Harvard Medical Library.

15. Peter W. F. Wilson et al., *New England Journal of Medicine*, 1985

16. Women's Health Initiative Investigators, "Risks and Benefits of Estrogen plus Progestin in Healthy Postmenopausal Women," *Journal of the American Medical Association*, July 17, 2002.

17. W. B. Kannel and P. Sorlie, "Some Health Benefits of Physical Activity: The Framingham Study," *Archives of Internal Medicine*, August 1979.

18. H. B. Hubert et al., "Obesity as an Independent Risk Factor for Cardio-

vascular Disease: A 26-Year Follow-Up of Participants in the Framingham Heart Study," *Circulation*, May 1983.

19. P. A. Wolf, R. D. Abbott, and W. D. Kannel, "Atrial Fibrillation as an Independent Risk Factor for Stroke: The Framingham Study," *Stroke*, August 1991.

20. T. Gordon et al., "High Density Lipoprotein as a Protective Factor Against Coronary Heart Disease: The Framingham Study," *American Journal of Medicine*, May 1977.

21. Peter W. F. Wilson et al., "Prediction of Coronary Heart Disease Using Risk Factor Categories," *Circulation*, May 12, 1998.

22. Claude Lenfant, "Preface," *Medical Milestones from the National Heart, Lung, and Blood Institute's Framhingham Heart Study*, ed. Daniel Levy 1999, p. xiii.

NINE: COMING OF AGE IN A HIGH-TECHNOLOGY WORLD

1. Interview, December 2003.

2. Interview, December 2003.

3. P. M. Okin et al, "Heart Rate Adjustment of Exercise-Induced ST Segment Depression: Improved Risk Stratification in the Framingham Offspring Study," *Circulation*, March 1991.

4. M. Bikkina, M. G. Larson, and D. Levy, "Prognostic Implications of Asymptomatic Ventricular Arrhythmias: The Framingham Heart Study," *Annals of Internal Medicine*, December 15, 1992.

5. D. Levy et al., "Echocardiographically Detected Left Ventricular Hypertrophy: Prevalence and Risk Factors," *Annals of Internal Medicine*, January 1988.

6. M. S. Lauer, K. M. Anderson, and D. Levy, "Influence of Contemporary Versus 30-Year Blood Pressure Levels on Left Ventricular Mass and Geometry: The Framingham Heart Study," *Journal of the American College of Cardiology*, November 1, 1991.

7. D. Levy et al., "Prognostic Implications of Echocardiographically Determined Left Ventricular Mass in the Framingham Heart Study," *New England Journal of Medicine*, May 31, 1990.

8. R. S. Vasan et al., "Left Ventricular Dilation and the Risk of Congestive Heart Failure in People Without Myocardial Infarction," *New England Journal of Medicine*, May 8, 1997.

9. S. M. Vaziri et al., "Echocardiographic Predictors of Nonrheumatic Atrial Fibrillation: The Framingham Heart Study," *Circulation*, February 1994.

10. M. Bikkina et al., "Left Ventricular Mass and Risk of Stroke in an Elderly Cohort: The Framingham Heart Study," *Journal of the American Medical Association*, July 6, 1994.

11. Memos from Daniel McGee, June 30 and December 20, 1983.

TEN: BLOOD PRESSURE:
MORE THAN JUST A NUMBERS GAME

1. T. R. Dawber, F. E. Moore, and G. V. Mann, "Coronary Heart Disease in the Framingham Study," *American Journal of Public Health,* April 1957.

2. Interviews with Edward Freis, January 2001–November 2003.

3. "The Seventh Report of the Joint National Committee on Prevention, Detection, Evaluation, and Treatment of High Blood Pressure," U.S. Department of Health and Human Services, NIH, NHLBI May 2003.

4. Julius H. Comroe, *Exploring the Heart* (Toronto: George J. McLeod, 1983), p. 216.

5. P. B. Beeson and W. McDermott, *Cecil-Loeb Textbook of Medicine,* 12th ed. (Philadelphia: Saunders, 1967), pp. 665–66.

6. Edward D. Freis, "Treatment of Hypertension with Chlorothiazide," *Journal of the American Medical Association,* January 10, 1959.

7. Richard Doll, "Controlled Trials: The 1948 Watershed," *British Medical Journal,* October 31, 1998.

8. Ibid.

9. Marcia Meldrum, " 'A Calculated Risk': The Salk Polio Vaccine Field Trials of 1954," *British Medical Journal,* October 31, 1998.

10. William Goldring and Herbert Chasis, "Antihypertensive Drug Therapy: An Appraisal," *Archives of Internal Medicine* 1965.

11. Veterans Administration Cooperative Study Group on Antihypertensive Agents, "Effects of Treatment on Morbidity in Hypertension," *Journal of the American Medical Association,* December 11, 1967.

12. E. D. Freis, "Reminiscences of the Veterans Administration Trial of the Treatment of Hypertension," *Hypertension,* October 1990.

13. N. E. Lassen, "Epidemiologic Assessment of the Role of Blood Pressure in Stroke," *Journal of the American Medical Association,* October 1996.

14. A. Kagan et al., "Blood Pressure and Its Relation to Coronary Heart Disease in the Framingham Study," *Hypertension,* July 1959.

15. Veterans Administration Cooperative Study Group on Antihypertensive Agents, "Effects of Treatment on Morbidity in Hypertension: II. Results in Patients with Diastolic Blood Pressure Averaging 90 through 114 mm Hg," *Journal of the American Medical Association,* August 17, 1970.

16. Interview with Edward Freis, January 2001.

17. William B. Kannel et al., "Epidemiologic Assessment of the Role of Blood Pressure in Stroke: The Framingham Study," *Journal of the American Medical Association,* October 12, 1970.

18. SHEP Research Group, "Systolic Hypertension in the Elderly Program," *Hypertension,* March 17, 1991.

19. ALLHAT Collaborative Research Group, "Antihypertensive and Lipid-Lowering Treatment to Prevent Heart Attack Trial," *Journal of the American Medical Association,* December 18, 2002.

20. Claude Lenfant, director, NHLBI, NIH news release, December 17, 2002.

21. Menard Gertler and Paul D. White, *Coronary Heart Disease in Young Adults* (Cambridge: Harvard University Press, 1954), 8.

22. Ibid.

23. Ibid.

24. ALLHAT, "Antihypertensive and Lipid-Lowering Treatment."

25. "The Seventh Report of the Joint National Committee."

26. D. E. Morisky et al., "Five-Year Blood Pressure Control and Mortality Following Health Education for Hypertensive Patients," *American Journal of Public Health,* February 1983.

27. R. S. Vasan et al., "Residual Lifetime Risk for Developing Hypertension in Middle-aged Women and Men: The Framingham Heart Study." *Journal of the American Medical Association,* February 2002.

28. Interview, March 25, 2002

ELEVEN: THE WAGES OF SIN

1. Donald S. Fredrickson, "Phenotyping: On Reaching Base Camp," *Circulation,* April 1993.

2. Interview with Aram V. Chobanian, dean, Boston University School of Medicine, April 2001.

3. W. F. Enos, "Coronary Disease among United States Soldiers Killed in Action in Korea: A Preliminary Report," *Journal of the American Medical Association* July 18, 1953.

4. Interview with Michael Brown, May 2002.

5. Michael Brown, "How Genes Control Cholesterol," lecture delivered in late 1990s.

6. P. Roy Vagelos, "Are Prescription Drug Prices High?" *Science,* May 24, 1991.

7. Ibid.

8. Interview with Jonathan Tobert, May 2002.

9. "Girl, 6, Undergoes Dual Transplant," *New York Times,* February 15, 1984.

10. Sheldon H. Gottlieb, "Stormie Jones: A Six-Year-Old Heroine," *American Diabetes Association,* December 1, 2001.

11. Vagelos, "Are Prescription Drug Prices High?"

12. Ibid.

13. Scandanavian Simvastatin Survival Study Group, "Randomised Trial of Cholesterol Lowering in 4,444 Patients with Coronary Heart Disease: The Scandinavian Simvastatin Survival Study (4S)," *Lancet,* November 19, 1994.

14. Shepherd J, Cobbe SM, Ford I, Isles CG, Lorimer AR, MacFarlane PW, McKillop JH, Packard CJ. Prevention of coronary heart disease with pravastatin in men with hypercholesterolemia. West of Scotland Coronary Preven-

tion Study Group. 1995 Nov 16; *New England Journal of Medicine,* 333 (20): 1301–7.

15. Downs JR, Clearfield M, Weis S, Whitney E, Shapiro DR, Beere PA, Langendorfer A, Stein EA, Kruyer W, Gotto AM Jr. Primary prevention of acute coronary events with lovastatin in men and women with average cholesterol levels: results of AFCAPS/TexCAPS. Air Force/Texas Coronary Atherosclerosis Prevention Study. *JAMA.* 1998 May 27; 279 (20): 1615–22.

16. Collins R, Armitage J, Parish S, Sleigh P, Peto R; Heart Protection Study Collaborative Group. MRC/BHF Heart Protection Study of cholesterol-lowering with simvastatin in 5963 people with diabetes: a randomised placebo-controlled trial. *Lancet.* 2003 Jun 14; 361 (9374): 2005–16.

17. Interview with Sharyn Weir, 1999.

18. Christopher P. Cannon et al., "Comparison of Intensive and Moderate Lipid Lowering with Statins After Acute Coronary Syndromes," *New England Journal of Medicine,* April 8, 2004.

TWELVE: RENEGADE ON THE TRAIL
OF THE UNKNOWN

1. J. Selhub et al., "Association Between Plasma Homocysteine Concentrations and Extracranial Carotid-Artery Stenosis," *New England Journal of Medicine,* February 2, 1995.

2. Interview with Kilmer McCully, December 2003.

3. Interview with Kilmer McCully, April 2001.

4. M. J. Stampfer et al., "A Prospective Study of Plasma Homocysteine and Risk of Myocardial Infarction in U.S. Physicians," *Journal of the American Medical Association,* August 1992.

5. Kilmer McCully, "The Significance of Wheat in the Dakota Territory, Human Evolution, Civilization, and Degenerative Diseases," *Perspectives in Biology and Medicine,* Winter 2001.

6. Kilmer McCully, "Homocysteine Theory of Arteriosclerosis: Development and Current Status," *Atherosclerosis Reviews,* Vol. 11, ed. by A. M. Gotto Jr. and R. Paoletti (New York: Raven Press, 1983).

7. Ibid.

8. Kilmer McCully, "Vascular Pathology of Homocysteinemia: Implications for the Pathogenesis of Arteriosclerosis," *American Journal of Pathology,* July 1969.

9. Guido Schnyder et al., "Decreased Rate of Coronary Restenosis after Lowering of Plasma Homocysteine Levels," *New England Journal of Medicine,* November 29, 2001.

10. James F. Toole et al., "Lowering Homocysteine in Patients with Ischemic Stroke to Prevent Recurrent Stroke, Myocardial Infarction, and Death: The Vitamin Intervention for Stroke Prevention (VISP) Randomized Controlled Trial," *Journal of the American Medical Association,* February 4, 2004.

11. H. A. Schroeder, "Losses of Vitamins and Trace Minerals Resulting from Processing and Preservation of Foods," *American Journal of Clinical Nutrition,* May 24, 1971.

THIRTEEN: THE LIFESTYLE REVOLUTION

1. Ancel Keys and Margaret Keys, *Eat Well and Stay Well* (Garden City, N.Y.: Doubleday, 1959).

2. Henry Blackburn, "Ancel Keys," essay for University of Minnesota, March 20, 2002, http://mbbnet.umn.edu/firsts/blackburn_h.html

3. William Hoffman, "Meet Monsieur Cholesterol," *University of Minnesota Update,* Winter 1979.

4. "The Fat of the Land," *Time,* January 13, 1961.

5. Hoffman, "Meet Monsieur Cholesterol."

6. Ibid.

7. University of Minnesota, Gateway Heritage Gallery, Minneapolis, Winter 1999.

8. Hoffman, "Meet Monsieur Cholesterol."

9. Ibid.

10. Ancel Keys, *Coronary Heart Disease in Seven Countries,* American Heart Association Monograph No. 29, 1970.

11. Henry Blackburn, *On the Trail of Heart Attacks in Seven Countries* (Middleborough, Mass.: Country Press, 1995).

12. Ibid.

13. Keys, *Coronary Heart Disease.*

14. Interview with Jeremiah Stamler, March 2002.

15. Ibid.

16. Ibid.

17. Interview with Oglesby Paul, January 2002.

18. Interview with Henry Blackburn, March 2003.

19. *Food for Your Heart: A Manual for Patient and Physician,* Department of Nutrition, Harvard School of Public Health, Harvard University, Cambridge.

20. *What We Know about Diet and Heart Disease,* American Heart Association, 1961.

21. Interview with Jeremiah Stamler, March 2002.

22. U.S. Senate Select Committee on Nutrition and Human Needs, "1977 Dietary Goals for the United States," 1977.

23. Ibid.

24. H. B. Hubert et al., "Obesity as an Independent Risk Factor for Cardiovascular Disease: A 26-Year Follow-Up of Participants in the Framingham Heart Study," *Circulation,* May 1983.

25. Michael L. Dansinger, press conference, American Heart Association Scientific Sessions, November 2003.

26. Antonia Trichopoulou et al., "Adherence to a Mediterranean Diet and

Survival in a Greek Population," *New England Journal of Medicine,* June 26, 2003.

27. U.S. Department of Health and Human Services, "Overweight and Obesity: A Major Public Health Issue," Prevention Report, 2001.

28. National Health and Nutrition Examination Survey, 1999–2000.

29. Blackburn, *On the Trail of Heart Attacks in Seven Countries,* p. 145.

30. Centers for Disease Control and Prevention, National Center for Health Statistics, Health, United States, 2000.

31. National Commission on Egg Nutrition v. FTC, 570 F.2d 157 (7th Cir. 1977).

32. USDA Center for Nutrition Policy and Promotion.

33. "Is Total Fat Consumption Really Decreasing?" USDA Center for Nutrition Policy and Promotion, April 1998.

34. USDA, Food Surveys Research Group, 1998.

35. Ibid.

36. Hubert et al., "Obesity as an Independent Risk Factor for Cardiovascular Disease."

37. Thomas M. Loftus et al., "Reduced Food Intake and Body Weight in Mice Treated with Fatty Acid Synthase Inhibitors," *Science,* June 30, 2000.

38. Interview with Ralph Paffenbarger, June 2002.

39. Interview with Oglesby Paul, January 2002.

40. R. S. Paffenbarger, Jr., A. L. Wing, and R. T. Hyde, "Physical Activity as an Index of Heart Attack Risk in College Alumni," *American Journal of Epidemiology,* September 1978.

41. Interview with Ralph Paffenbarger, January 2002.

42. Kyle McInnis, "A Heart Strengthening Pace: Brisk but Comfortable," American Heart Association meeting report, November 2003.

43. "Physical Activity among Adults: United States, 2000," National Center for Health Statistics, Advance Data, 2003.

FOURTEEN: DEADLY ADDICTION

1. Richard Kluger, *Ashes to Ashes: America's Hundred-Year Cigarette War, the Public Health and the Unabashed Triumph of Philip Morris* (New York: Knopf, 1997).

2. J. T. Doyle et al., "Cigarette Smoking and Coronary Heart Disease: Combined Experience of the Albany and Framingham Studies," *New England Journal of Medicine,* April 19, 1962.

3. Centers for Disease Control and Prevention, "MMWR Highlights—Cigarette Smoking Among Adults—United States, 2001," October 10, 2003.

4. "First Surgeon General's Report on Smoking and Health," January 11, 1964.

5. Interviews with William Kannel, September 1998–July 2003.

6. Gene Borio, "Tobacco Timeline," *Tobacco News and Information,* 2001 (www.tobacco.org).

7. John R. Bumgarner, *The Health of Presidents* (Jefferson, N.C.: McFarland, 1994), pp. 225–33.

8. Winea J. Simpson, "A Preliminary Report on Cigarette Smoking and the Incidence of Prematurity," *American Journal of Obstetrics and Gynecology,* April 1957.

9. William Kannel, "Tobacco and Cardiovascular Disease," Rough Draft No. 1, September 18, 1963.

10. Centers for Disease Control and Prevention, "Age of Initiation of Cigarette Smoking—United States, 1991."

11. Centers for Disease Control and Prevention, "History of the 1964 Surgeon General's Report on Smoking and Health," compiled by the Office on Smoking and Health, January 1994–July 1996.

12. William H. Stewart, Surgeon General, Second Surgeon General's Report, "The Health Consequences of Smoking: A Public Health Service Review," 1967.

13. Interviews with William Kannel, 1998–2003.

14. *Tobacco News and Information.*

15. Interviews with William Castelli, 1998–2003.

16. NHLBI, Framingham Archives.

17. Interview with Robert Garrison, March 2003.

18. W. P. Castelli et al., "The Filter Cigarette and Coronary Heart Disease: The Framingham Study," *Lancet,* July 8, 1981.

19. P. A. Wolf et al., "Cigarette Smoking as a Risk Factor for Stroke: The Framingham Study," *Journal of the American Medical Association,* February 19, 1988.

20. Centers for Disease Control and Prevention, "MMWR Highlights."

21. CDC Office of Communication, "Teen Smoking Rates Decline Significantly," press release, May 16, 2002.

22. American Heart Association, "2000 Heart and Stroke Statistical Update," December 30, 1999.

FIFTEEN: SPREADING THE WORD

1. National Hot Dog and Sausage Council, Vital Hot Dog Statistics, Published Facts.

SIXTEEN: A FIFTY-YEAR PROMISE UNDER ATTACK

1. William P. Castelli and Philip A. Wolf, letter to "Friend of Framingham or Study Participant," April 27, 2000.

2. Interview with David Lampe, November 2003.

3. Allison Connolly, "Heart Study Spinoff Raises Funds Questions," *Boston Business Journal,* May 19–25, 2000.

4. Interview with Fred Ledley, November 2003.

5. Fred Ledley document, "Putting a Human Face on Genomics."

6. Interview with Arthur Caplan, July 2003.

7. Castelli and Wolf, letter to "Friend of Framingham or Study Participant."

8. Interveiw with Castelli, December 2003.

9. Dave Granlund, "Data for Profit Scheme," *MetroWest Daily News*, August 13, 2000.

10. Sandra N. Fitts, "Heart Study Data in Good Hands," *MetroWest Daily News*, July 13, 2000.

11. Daniel J. Liberatore, Letters to the Editor, "Release of Heart Study Data Would Be Breach of Trust," *MetroWest Daily News*, July 23, 2000.

12. Interview with Ledley, November 2003.

13. Jay J. Lander, letter to Aram Chobanian, Dean, Boston University School of Medicine; to Claude Lenfant, director, NHLBI; and to Fred Ledley, CEO, Framingham Genomic Medicine, June 28, 2000.

SEVENTEEN: THE GENETIC REVOLUTION

1. L. Atwood et al., "Genomewide Linkage Analysis of Body Mass Index across 28 Years of the Framingham Heart Study," *American Journal of Human Genetics*, November 2002

2. D. Levy et al. "Evidence for a Blood Pressure Gene on Chromosome 17: Genome Scan Results for Longitudinal Blood Pressure Phenotypes in Subjects from the Framingham Heart Study," *Hypertension*, October 2000.

3. A. M. Shearman et al., "Association Between Estrogen Receptor Alpha Gene Variation and Cardiovascular Disease," *Journal of the American Medical Association*, January 14, 2003.

EPILOGUE

1. Associated Press, "Summary of Cheney's Heart Problems," June 29, 2001.

2. Shankar Vedantam, "For Cheney's Heart, an Electrical Test; Implant Would Keep Beat Steady," *Washington Post*, June 30, 2001.

3. American Heart Association, "2000 Heart and Stroke Statistical Update," December 30, 1999.

INDEX

A Note about the Authors

Daniel Levy, M.D., is the current director of the Framingham Heart Study.

Susan Brink is a Senior Writer covering medicine for *U.S. News & World Report.*

A Note on the Type

This book was set in Galliard, a typeface drawn by Matthew Carter for the Mergenthaler Linotype Company in 1978. Carter, one of the foremost type designers of the twentieth century, studied and worked with historic hand-cut punches before designing typefaces for Linotype, film, and digital composition. He based his Galliard design on sixteenth-century types by Robert Granjon. Galliard has the classic feel of the old Granjon types as well as a vibrant, dashing quality that marks it as a contemporary typeface and makes its name so apt.

Composed by Creative Graphics, Inc., Allentown, Pennsylvania
Printed and bound by Berryville Graphics, Berryville, Virginia
Designed by Robert C. Olsson